D0742643

Social Processes of Scientific Development

Social Processes of Scientific Development

Edited by

Richard Whitley

Routledge & Kegan Paul
London and Boston

First published in 1974
by Routledge & Kegan Paul Ltd
Broadway House, 68–74 Carter Lane,
London EC4V 5EL and
9 Park Street,
Boston, Mass. 02108, U.S.A.
Printed in Great Britain by
W & J Mackay Limited, Chatham

ISBN 0 7100 7705 X

Contents

List of Contributors

Stuart S. Blume, M.A., Dr Phil., Research Officer, Civil Service College, London, and Consultant, Directorate for Scientific Affairs, OECD, Paris

Norbert Elias, Dr Phil., Former Professor of Sociology, University of Ghana; Reader in Sociology, Leicester University; Visiting Professor, University of Amsterdam; Visiting Professor, University of Konstanz

János Farkas, Dr Sc., Deputy Director of the Institute of Sociology, Hungarian Academy of Sciences

Thorvald Gran, Ph.D., Lecturer in Public Administration, Institute of Sociology, University of Bergen

Rolf Klima, Cand. Phil., Research and Teaching Assistant, Faculty of Sociology, University of Bielefeld

Zdislaw Kowalewski, Dr Phil., Director of Centre for the Social Function of Science, Institute of the History of Science, Polish Academy of Sciences

Cornelis J. Lammers, Ph.D., Professor of Sociology of Organization, University of Leyden, Fellow of Netherlands Institute for Advanced Studies, Wassenaar

Albert Mok, Ph.D., Professor of Organizational Sociology, University of Antwerp

Benjamin Nelson, Professor of Sociology and History, Graduate Faculty, New School for Social Research, New York

Hilary Rose, B.A., Lecturer in Sociology, Dept. of Social Administration, London School of Economics and Political Science

Steven Rose, Ph.D., Professor of Biology, The Open University

Wouter van Rossum, Soc. drs., Lecturer in Methods and Techniques of Social Research, Dept. of Sociology, University of Amsterdam

Ruth Sinclair, B.Sc.(Econ.), Research Fellow, Dept. of Social Sciences, University of Loughborough

Peter Weingart, Dr Phil., Professor of Sociology of Science, University of Bielefeld

Anne Westerdiep, B.A., Researcher at the Institute of Applied Sociology, Nijmegen, Netherlands

Richard Whitley, M.A., Lecturer in Sociology, Manchester Business School, University of Manchester

Dorothy S. Zinberg, Ph.D., Lecturer in Sociology, Harvard University

Acknowledgments

I am very grateful to all contributors to this volume for their promptness in revising their chapters and toleration of editorial alterations and condensation. I was greatly assisted in the editorial work by Donna Andrews who ensured that the authors' meaning was respected. Carolyn Taylor carried out the clerical work with inestimable competence and quickness. Finally, I am indebted to Albert Mok and Benjamin Nelson for acting as my editorial advisers.

Introduction

Richard Whitley

The papers in this book arose out of a conference, held in London in September 1972, of the International Sociological Association's Research Committee on the Sociology of Science, jointly sponsored by the British Sociological Association. One of the main purposes of the meeting was to discuss new developments and directions in the field which had occurred since the meeting held at the World Congress of Sociology at Varna, Bulgaria, two years earlier. In this introduction, I will briefly discuss some of these developments, their connection with previous work and possible future directions indicated by participants at the conference as well as some implications for sociology.

Implications for the sociology of science

The common thread throughout the conference was an agreement that science was a cultural activity amenable to sociological analysis. Science in this sense included the activities of scientists, the cognitive structures they produce, develop and alter as well as their social relations. Despite differences in epistemological outlooks, it was agreed that the sociology of science had to concern itself with the content of a science, the development of that content, the different modes of development in different sciences and the connections between scientific developments and cultural and institutional factors. The epistemological issues raised by a sociological analysis of scientific developments were acknowledged, but not seen as precluding such analysis (cf. Lakatos, 1971). This consensus represented a considerable change in the dominant concerns of the field. Most research in the sociology of science had concentrated on hypostatized 'norms' of science, reward allocation procedures among scientists, communication structures of scientific groups and the social organization of science. This tradition has been extensively analysed and does not require further discussion (cf. Barnes and Dolby, 1970; King, 1971; Martins, 1972; Whitley, 1972). Generally speaking, sociologists did not examine the content of scientists' work or the development of scientific ideas. The history and philosophy of science tended to be seen as having pre-empted these areas.

A major factor in this reorientation of the sociology of science was the publication of Kuhn's *The Structure of Scientific Revolutions* (1962; 1970) which can be seen as the culmination of growing discontent with the logical empiricist school in the philosophy of science and of the Koyréan 'revolution' in the history of science. Coterminous with this was the development of a new variant in the Popperian school of the philosophy of science which is associated with Feyerabend and Lakatos although general awareness of this movement did not occur until the publication of the Kuhn–Popper debate in 1970 (Lakatos and Musgrave, 1970). By allowing for a degree of epistemological irrationalism among scientists, Kuhn opened the door for a sociological analysis of scientific development. The denial of the inevitability of scientific progress by means of applying *the* scientific method raised the possibility of scientific error and so the need for understanding such error. This view suggested that traditional philosophy and history of science had not, after all, pre-empted the sociological study of scientific ideas.

While epistemological relativism had been discussed before, notably by Polanyi (1958), Kuhn elaborated a concept which combined cognitive aspects with sociological ones. The term 'paradigm', although susceptible to many interpretations (cf. Masterman, 1970), provided sociologists with a category for exploring scientific ideas in relation to scientific communities. Also, of course, it formed part of a model of scientific development which could be applied directly to cognitive enterprises so that they could be compared and reasons for differences in development propounded. In other words, as well as allowing for the possibility of a sociology of scientific development Kuhn outlined a framework for such a sociology and so provided the basis for a radical change in the sociology of science.

However, direct application of the Kuhnian model has certain drawbacks and many modifications have been offered. The chief difficulties concern the monistic nature of the 'paradigm', at least in its first formulation, and the unitary model of scientific development. As Elias points out, much philosophy of science is really the philosophy of physics and a not even up to date physics at that; Kuhn, unfortunately, can be tarred with the same brush for he talks of a 'mature' science following his model and the major example of 'maturity' is held to be physics. Indeed, it is possible that 'immature' sciences are not really sciences at all (cf. Martins, 1972). Thus, a unitary model becomes a demarcation criterion for distinguishing science from non-science. The emphasis at the conference on the need for comparative studies of the development of different sciences requiring commonality of categories and the existence of a plurality of possible modes of development is reflected in many of the chapters (Nelson, Elias, Weingart, Whitley, Kowalewski, Blume). A further modification is the move away from regarding cognitive structures in each science as monistic and fully integrated. It is increasingly being recognized that a science does not consist of one logically closed, fully connected cognitive structure which is either stable or in a 'revolutionary' situation. Not only do sciences differ with regard to the degree of integration, but different types of cognitive structure within each science can be differentiated and their degree of cognitive integration may vary (Weingart, Whitley, Lammers).

These two modifications result in a need for elaboration of categories for cognitive structures within sciences and their development. While the attempts at such elaboration published here are necessarily schematic and unproven in comparative studies, the growing number of case studies of particular scientific developments should enable their fruitfulness to be assessed and modifications and alternatives suggested. The main point is that distinct levels of cognitive structure have to be differentiated, their interconnections understood and modes of development identified. Concomitant with the formulation of cognitive categories, of course, is the elaboration of social groups and structures so that the cognitive and social developments can be understood as interdependent (van Rossum). For a real comparative sociology of the sciences to be possible, formulation of categories which are common across sciences is essential; equally it is necessary to be clear as to the sociological import of these categories and the nature of the sociological problem under investigation. Categories used by different scientists often do not correspond and, *a fortiori*, do not overlap with the sociologists'. Reliance on scientists' 'common sense' interpretation of categories does not guarantee reliability of responses across fields.

It is important to recognize that while rejection of the Kuhnian unitary model of scientific development implies that development is not equated with necessary, progressive evolution (cf. Toulmin, 1972), this view need not lead to relativism. Insistence on the pluralism of the sciences and their ways of development is not the same as asserting the irrelevance of epistemological criteria for a sociology of the sciences. Agnosticism towards theories of knowledge does imply epistemological relativism though, as Barnes (1972) acknowledges. If, for example, it is suggested that sociologists should concentrate on studying scientists' beliefs and cognitions irrespective of the validity of the standards used to evaluate them, the implication is that any one set of standards is as epistemologically valid as any other. The result of advocating such relativism is the lack of any clear indication as to how competing sociological rationalities are to be compared.

In selecting a scientific development as sociologically problematic, certain assumptions are made about what is 'normal' and 'abnormal', hence, in need of explanation. Each science selects and describes certain phenomena as relevant and as problematic in terms of its theoretical structures. These structures determine, to varying degrees, what is expected and so non-problematic and what is unexpected, but describable and so requires understanding. Sociology is no exception to this and imputes a rationality to social phenomena which determines objects for investigation and criteria for the evaluation of solutions to the problems raised by such objects. While theoretical structures are rarely cohesive and elaborated, they are implicit in every sociological study. Criteria of 'normal' behaviour necessarily follow from the designation of some phenomenon as sociologically problematic. Every investigation delineates a problem and so implies some standard of rationality in terms of which action is 'reasonable'. In the sociology of science, this might take the form of the Kuhnian view where a scientist would be 'rational' if, as a result of intensive socialization, he concentrates on extending the existing dogma and faithfully obeys the dictates of the élite. Innovation would, then, be sociologically problematic, but it would not be possible to say anything about the epistemological validity of any assertion. The problem with this approach is that in denying the possibility of evaluating competing sets of standards in natural science there is no way of preferring one sociological rationality to another. What is 'normal' to one sociologist may be 'abnormal' to another and there is no way under radical relativism in which this conflict can be resolved. Nor, of course, is there any way competing explanations can be evaluated assuming some commonality of problem solution and definition. In other words, sociological rationality and epistemological rationality must overlap at some point if we are to claim that a progressive sociology of science is possible. Unless the relativists are willing to admit to total subjectivism, in fact, some overlap is necessarily implied by the designation of some phenomenon as problematic. The imputation of rationality and associated standards for delineating a problem imply some way in which the selection of that rationality can be justified. That is, the sociologist, in designating a phenomenon as sociologically problematic, is invoking, however tenuously, a set of common standards according to which his problem and solution can be discussed and assessed. Either the relativist asserts the total irrationality of his own problem solution in other words, or he is able to justify that selection in terms of some standards which imply a concept of correct sociological work and so 'science' in some sense. Whatever these standards may be, some are necessarily invoked once a particular scientific development is selected as problematic and so the sociologist cannot remain agnostic towards epistemological issues. It would seem peculiar to allow sociologists to evaluate problem selection standards such that choices can be rationally justified but not to extend that privilege to physicists. In urging recognition of cognitive pluralism, then, it is not being suggested that there are no criteria for comparing sciences, but that the differentiation of the sciences must be considered and differences in the nature of the objects studied allowed for.

The Kuhnian tradition, and its modifications, elaborations and alterations, has tended to concentrate on the relations between scientific communities and cognitive structures rather than examine the wider social environment or science as a civilizational product (Nelson). Similarly, the application of exchange theory to the study of science has generally been restricted to the reward system in scientific communities. While the market analogy may have derived from commodity capitalism, it has not provided a link between the evaluation of scientific products and standards elsewhere in society. Indeed, it is not entirely clear how firm the analogy is supposed to be since scientific papers and ideas are not treated as commodities for exchange, but as products for use. As Klima points out, scientists allocate rewards, such as recognition, on the basis of the importance of the problem a particular idea or solution is relevant to and to the extent that it 'solves' that problem. By tying the allocation of rewards to incongruity theory, Klima has provided a cognitive foundation for the operation of exchange and has connected the cognitive decisions taken by scientists with social judgments and extra-scientific considerations. The importance of the incongruity reduced by a particular idea is socially determinate and can, thus, be influenced by societal values although, of course, it does not necessarily follow that scientists will always concur with society's priorities. They may ignore social and economic pressures to reorder their priorities or merely pay lip service to extra-scientific views. The criteria invoked in deciding which problems are important in a particular science may be purely local—in terms of the colleague community—or both local and societal (Mok and Westerdiep), but not, presumably, as long as the scientific identity is preserved, purely societal.

The social conditions under which local or societal criteria predominate are examined by Lammers in a discussion of the differential patterns of development shown by the natural and social sciences. The institutional approach to the science-society relation has often been separated from the cultural approach which examines connections between metaphysical ideas in a science and the dominant ideas of a particular society, but Lammers analyses the connections between cultural, 'lay' images and the institutional arrangements surrounding a science in such a way as to explicate differences in the elaboration of theoretical frameworks. Similarly, the Roses connect the dominant socio-economic structure to particular ideas about reality in the dominant culture and explore the impact of these ideas in the development of a number of related sciences. The artificial distinction between the history of ideas and social reliationships in the development of a science is denied.

The Marxist tradition, of course, has never considered the 'internal-external' distinction in the history of science (cf. Kuhn, 1968) valid. Relations between science and society have always included actual scientific developments and the separation of ideas, knowledge and scientists' social behaviour is seen as unnecessary (Gran). The study of science as a total, social phenomenon is emphasized by Farkas in his discussion of the science of science. The concrete analysis of scientific development in society is still, however, relatively embryonic as, indeed, is the understanding of the development of scientists as discussed by Zinberg. Detailed consideration of the way different levels of various sciences have developed in particular cultural and institutional contexts is required, as well as inquiry into the links between types of science or 'sciencing' and cultural world views. The admission of a multiplicity of relatively independent cognitive structures in science necessitates study of each in relation to developments elsewhere. The assumption that understanding the relation between a metaphysical postulate of a science and dominant cultural images automatically leads to understanding of how all aspects of the science developed is no longer tenable.

Implications for sociology

In many respects, the papers published here do little more than indicate a large area for future work. In so doing, however, they represent a redefinition of the sociology of science and one, moreover, which is important for sociology as a discipline in that it insists on combining the study of traditional social indicators with detailed consideration of the nature and development of the particular social activity being analysed. The sociologist wishing to understand science is now examining the detailed actions of scientists in the laboratory, often with the active collaboration of working scientists, individual and group judgments concerning interpretation of results and problem solutions and the development of ideas as well as the journal structure of a science, the educational stratification system, the pattern of citations and social networks. If this can be done in studying science, surely it is even more appropriate in looking at other, less autonomous, social phenomena.

In carrying out this task the sociologist is forced to reflect, rather more perhaps than with other phenomena, on the nature of the sociological problem. This is partly because historians and philosophers have been concerned with science for some time but primarily because science itself is intensely self-conscious and has an elaborate system of reconstructions of its ideas which is often far superior to any mere sociological intellectual edifice. The 'common sense reality' of the scientist is seen by many as of a qualitative different order to that of a school teacher, policeman or car assembly worker to the extent that it is not taken as sociologically problematic. Implicitly or explicitly the scientist's definition of his work and results is accepted and, hence, does not constitute a sociological phenomenon to be understood. While in other areas of sociology, the imposition of the sociologist's meaning on a social situation is under attack, in the sociology of science the reverse is the case. Here, the sociologist starts from the 'official' history, or reconstruction (cf. Lakatos, 1971), and has to formulate his problem as one of understanding that history in terms of his own reconstruction. In most areas of social life the sociologist has to reconstruct reality *ab novo*, from a variety of cues, and does not have a fully rationalized history to contend with. He is able to flatter himself that he is the possessor of the real 'story', since there does not appear to be any realistic, systematic alternative. In science, and to a lesser extent in other cultural activities, the sociologist has to come to terms with an elaborate rationalization whose existence implies that the sociologist is redundant or, at best, only fit to pick up the crumbs from the philosophers' table (Lakatos, 1971). In wrestling with this system of ideas, the sociologist is forced to be much clearer about his objectives and procedures than if he had no competition. Much of the value of the sociology of science for sociology as a whole results from the necessity of dealing with major problems inherent in any sociological enterprise, but particularly acute in the study of science.

A further result of the reorientation of the sociology of science is that it directly leads to the development of the sociology of knowledge and the sociology of culture. Science being defined as a cultural activity and regarded as subject to sociological scrutiny suggests the extension of ideas in the sociology of science to the study of culture as a whole. While it may not be fruitful to go as far as Feyerabend (1967) does when he claims that art progresses and science does not, there seems little doubt that many of the categories and techniques being developed for analysing science can be used for understanding other cultural products. Especially, the pitfalls which are now a commonplace in the study of science should be avoidable in the sociology of art, etc. In general, it is to be hoped that the increased interest in the sociology of science is but the prelude to further work in the sociology of culture.

To a considerable extent the growth of work in the sociology of science is part of a general interest in power in society and a renewed search for the *primum mobile* of change. Science is seen as the exemplification of rationality which is coming to dominate social relations and one which, unlike practically all cultural activities hitherto, has developed a high degree of autonomy from the rest of society. While this view can result in hyperbole the manner in which the sciences and technologies have acquired and maintained their autonomy undoubtedly requires investigation, although here again care has to be exercised in boundary definition and the differentiated nature of cognitive structures recognized. The question of how science affects other social and cultural institutions similarly needs careful formulation and the whole area of the relations between culture and society is suggested here. As the sociology of science develops it can be expected to take an increasing interest in this area and should lead to greater understanding of these complex macro social phenomena, but at the moment, as with many other problems, work is in the embryonic stage.

Finally, it should be noted that most of the papers have had to be condensed for this book and that much of what was eliminated contained discussions of previous work. A certain minimum level of awareness, then, of the history and philosophy of science and of the Mertonian tradition in the sociology of science is assumed. Further discussion of earlier developments can be found in the references to this introduction and to the chapters in the book.

References

Barnes, S. B. (1972), 'Sociological explanation and natural science', *European Journal of Sociology*, 13, pp.373–91.

Barnes, S. B. and Dolby, R. G. A. (1970b), 'The scientific ethos: a deviant viewpoint', *European Journal of Sociology*, 11, pp.3–25.

Feyerabend, P. K. (1967), 'On the improvement of the sciences and the arts' in Cohen, R. S. and Wartofsky, M. (eds), *Boston Studies in the Philosophy of Science*, III, Dordrecht: Reidel.

King, M. (1971), 'Reason, tradition and the progressiveness of science', *History and Theory*, 10, 1, pp.3–32.

Kuhn, T. S. (1962), *The Structure of Scientific Revolutions*, Chicago University Press; 2nd ed. (with postscript), 1970.

Kuhn, T. S. (1968), 'History of science' in *The Encyclopaedia of the Social Sciences*, 2nd ed., New York: Macmillan.

Lakatos, I. (1971), 'History of science and its rational reconstructions' in Cohen, R. S. and Buck, R. (eds), *Boston Studies in the Philosophy of Science*, VIII, Dordrecht: Reidel.

Lakatos, I. and Musgrave, Alan E. (eds) (1970), *Criticism and the Growth of Knowledge*, Cambridge University Press.

Martins, H. (1972), 'The Kuhnian revolution and its implications for sociology' in Hanson, A. H., Nossiter, T., and Rokkan, S. (eds), *Imagination and Precision in Political Analysis: Essays in Honour of Peter Nettl*, London: Faber and Faber.

Masterman, M. (1970), 'The nature of a paradigm' in Lakatos, I. and Musgrave, Alan E. (eds), *Criticism and the Growth of Knowledge*, op. cit.

Polanyi, M. (1958), *Personal Knowledge*, London: Routledge & Kegan Paul.

Polanyi, M. (1964), *Science, Faith and Society*, Chicago University Press.

Toulmin, S. (1972), *Human Understanding*, London: Oxford University Press.
Whitley, Richard D. (1972), 'Black boxism and the sociology of science' in Halmos, P. (ed.), *The Sociology of Science*, Sociological Review Monograph No. 18, Keele University.

Perspectives on the Sociology of Science

1

On the shoulders of the giants of the *comparative* historical sociology of 'science' —in civilizational perspective*

Benjamin Nelson

Evidence mounts that we need to review our points of departure in the sociology of science.[1] The social-psychological, micro-sociological perspectives which have long governed the research done on scientific institutions and communities have now largely spent themselves.[2] The great issues which were the original setting of these inquiries are only rarely recalled today.[3]

The wider sociology of science, in the sense intended here, currently needs to be more clearly distinguished than it is from the more segmental issues with which it is too often identified to the detriment of the

* The original longer title provided a fuller sense of intended theoretical contexts. Specific reference was suggested to 'Maine, Durkheim, Mauss, Weber, the young Merton, Joseph Needham and other pioneers' dedicated to the study of 'science' and 'sciencing' in the wider settings of civilizational patterns and intercivilizational relations. The names of many—not all—of these other contributors are cited toward the end of the chapter and in the References below. An introduction to the present discussion can now be read in my review-essay on the recent reissue—with an important new Preface dated 1970—of Professor Merton's early classic, *Science, Technology, and Society in Seventeenth Century England* (cf. B. Nelson, 1972, in References).

[1] Proofs of the need for new approaches are offered in a series of papers, some of which are now going through the press. The present statement relates mainly to one aspect of the issue under discussion, namely, the *civilizational perspective* emphasized in the title. More is said about wider *societal* influences upon science and *societal* outcomes of science and technology in papers listed in the references under the present writer's name, e.g. 1968a, 1972, 1973.

[2] For a survey of this work see Cole and Cole (1967), Ben-David (1971).

[3] See, however, Merton's important new Preface dated 1970 to the recent reissue of doctoral dissertation (Merton, 1970, especially at pp.xii, xvii–xviii). Cf. the present writer's discussion (Nelson, 1972).

systematic progress of broader-gauged research. I refer to such familiar topics as: the interpersonal social relations among the social rankings of scientists in university departments or extra-academic research institutes or agencies or the fashions in the forms of citation in professional papers.

An indispensable step in attaining the objective here stressed is to renew the spirit and recollect the substance of the work done by the classical and contemporary pioneers of the comparative historical sociology of social and cultural process. I refer to such men as Maine (1861, 1871), Durkheim and Mauss (1963, 1972), Max Weber (1946, 1958), the young Robert Merton (1970) and Joseph Needham (1954) and others who have offered new horizons in special area studies. Although only a few of the aforementioned—notably Merton and Needham—place primary emphasis on the sociology of 'science', they are all acutely sensible to the need to find wider clues to the changes in the images and definition of realities. Extraordinarily acute insights remain to be recovered from their work.[1]

If we are to have sociology of science in the spirit required for our times of 'axial' conflicts in our horizons and sensibilities, we shall have to give renewed stress to the fact that 'science' refers pre-eminently to *interepochal* structures of consciousness, languages, modes of awareness, cognitive mappings, structures of rationales, agreed-upon procedures, symbolic technologies and so on (cf. Nelson, 1968b). In culture-areas marked by very comprehensive spreads—and equally comprehensive rejections—of rationalization in consciousness, sciences are languages of an extremely high degree of abstraction; decision-matrices are generally highly formalized mathematical or logical sets—so called nomological nets. Any effort to define the social factors and outcomes of science must do so in relation to the sum total of elements in the societal and civilizational setting in which one or another science locates. This means that the sociology of science must be historical and comparative. To link Durkheim and Maitland, we may say that the sociology of science is comparative and historical—and systematic—'or it is nothing'.

Purely retrospective assumptions invariably bar access to the ways in which *sciencing* occurs and 'sciences' are arranged in different places at different times. If our sights are only set to locate so-called precursors of present-day 'truths' we fall into the trap of writing separate histories of separate sciences as though these sciences were individuated from

[1] A paper of mine now in press (1973) explores these questions.

the start, consecutive in development, and unconnected with wider and deeper changes in the structures of existence, experience and expression.[1]

A large part of the confusion in this field results from the too often unexamined assumption that the present level and form of sciencing in the most advanced Western lands needs to constitute the criterion for the life and form of science in all other times and climes.

The contemporary situations in respect to the mixes in both the so-called 'West' and 'East' of the active elements and components of cultural complexes—the mixes of rational-non-rational, abstract-empirical, technology, magic, superstition—demand a wider understanding of the meanings and forms which sciences have taken in different settings than we possess today.[2] Robin Horton (1971) and a number of others have recently raised crucial questions on the ethnocentric character of our available definitions of science. Even though I cannot accept the peak assumptions and conclusion of Horton and others who adopt his view, I would agree that we need to look closely at the way in which sciencing occurs in settings of the widest variety. My reasons for stopping short of agreement with the full relativist position on 'sciencing' will have to be left for another occasion. Here it need only be noted that the link between these arguments and the largest cultural political questions of the present cannot be too strongly emphasized.

Again on this matter we will prove to have need of the clues afforded us by Maine, Durkheim, Mauss, the Durkheimians, the young Merton, Needham, and others.

In a series of papers I have written over the past decade I have been urging the need to stand on the shoulders of those giants who have applied wider comparative historical perspectives to the understanding of what I have called 'civilizational complexes and intercivilizational encounters' (cf. Nelson, 1973). Throughout this work I have sought to recover the rich ores mined by a number of writers now often forgotten, men who pioneered in the comparative historical study of religions, jurisprudence and forms of sensibility in civilizational perspective.

The most important spur which now beckons us forward today is the challenge of Joseph Needham. The issues which Needham has posed

[1] An approach to this position and emphasis will be found in Agassi (1963).
[2] Discussion on this point will be found in many papers, in Crombie (1963); Holton (1965), and other works which have touched upon the wider contexts of scientific development.

for the development of the comparative historical sociology of science are no less important than those implicit in the work of Max Weber. Indeed, our next major step in this field of studies will be the collation, comparison and contrast of the perspectives, data and hypotheses put forward by Needham and Weber in respect to the distinctive patterns, paces, outcomes of scientific development in the so-called 'East' and 'West' (see Needham, 1954, 1963, 1969).

It is the view of the present writer that all efforts to achieve this end will quickly lead us to acknowledge that so-far largely untapped resources left us by Maine, Durkheim, Mauss, the Durkheimians and Max Weber need to be exploited if we are to meet Needham's challenges—his implicit as well as explicit challenges.

An indispensable reference point of this analysis will be the study of factors working to promote and those working to retard the types of forgings of the types of universalities and universalizations necessary for the institutionalization[1] of innovation in 'advancement of science'.

In this connection there would be need to be as resourceful and precise as possible in spelling out the reciprocal—what some would call 'dialectical'—interanimations of the multiple universalizations occurring at the different levels of institutional and cultural structure and process. If the highest peaks of universalization are to be approached there have to be concurrent movements in the following directions:

(a) Bars to freedoms of entry into and exit from the communities of learners and participants and interpreters in the communities of discourse have to be lowered; inherited invidious dualisms have to be transcended.[2]

(b) Incentives to produce and distribute warranted knowledge, including new mapping and innovating procedures, have to be multiplied by deliberate policies.

(c) Blocks to the achievement of ever-higher levels of generality in the language structures—written and spoken—have to be surmounted. In this spirit we must attend to the thrusts to higher levels of abstraction

[1] Nelson (1973). It will be noted that the horizons of our stress on 'universalizations' range beyond the more restricted images of 'universalism' now accepted as a component of the ethos of the 'scientific community'. Most recently some of the larger aspects of the problem have received renewed emphasis by Merton. See his strong statement 'Insiders and Outsiders' (Merton, 1972).

[2] Critical clues to this issue may be found in Durkheim, especially (1915); Durkheim and Mauss (1963, 1972); and Weber (1946, 1958).

in the forming of special languages meant to serve as maps of terrain explored and remaining to be explored.[1]

There is no sociological issue which does not come under scrutiny when we become intent on tracing out the dialectical crossings of universalization processes and effects prerequisite to the spread of rationalized science. We no sooner confront Needham's challenges directly than we find ourselves tapping the resources of the classical pioneers in relating to the complex of critical issues relating to the passages to universalities in the terms of reference; the spread of universalization processes in the widenings of the communities of discourse; the expansions of opportunities for full participation in the community of learners, interpreters and teachers; the overcoming of particularistic inhibitions to the spread of communication and community; the encouragement of the 'individuals' intent upon advancing the exploration of every facet of their experience; the availability at every level of social and cultural differentiation of accepted tests of warranted assertibility.

Questions of this sort involve us in every aspect of the fortunes of social institutions and culture structures of 'science'.

Other notable points and corollaries which arise in this expanded view might be stated as follows:

(1) Comparative historical sociology of science must begin with the supposition that histories are multiple and that all sciencings occur in the course of histories and are themselves histories. There is nothing that requires us to suppose that any one place will serve as the permanent abode of any or all sciences. In the same way we dare not suppose that different sciences will forever have the same relative relation to one another or that groups or cadres linked to the sciences will share similar orientations at all points.

(2) We accomplish little when we treat actual histories as a reservoir of instances which go to prove one or another abstract proposition of structural-functional analysis or general systems theory. In some of their manifestations these approaches are less the roads of an advancing sociology of 'science' than surrogate faiths.

(3) To see 'sciencing' and sciences *in all the contexts of socio-cultural*

[1] Only a few sociologists have addressed themselves to this question which has been generally left to philosophers and mathematicians such as A. N. Whitehead, B. Russell, M. Klein, T. Danzig. We are in need of additional 'sociolinguistic' studies of higher level abstract languages. An exciting awareness of the thrust of these problems will be found in Chou En-Lai's statement on the reform of languages (Crozier, 1970, pp.128–31).

process and interaction, we have to adopt *civilizational perspectives* in the study of civilizational complexes and intercivilizational encounters.

I have discussed elsewhere the variable constitutions of civilizational complexes and the variable outcomes of intercivilizational encounters. Under this last rubric, I would call attention to the available research on the intercivilizational encounters of the twelfth and thirteenth centuries—involving societal settings, influences, and outcomes of the encounters of Chinese, Muslim, Hebrew, Byzantine and Western Christian encounters;[1] the encounters of China and the West in the sixteenth century and from the time of the Matteo Ricci mission to the expulsion of the Jesuits; the struggles over the new science in nineteenth century India; the conflicts over Western science and civilization in Tsarist Russia and the Soviet Union; the Nazi challenges to Western science.

We dare not close without noting that the stress throughout this paper has been on the *comparative historical sociology of science* IN CIVILIZATIONAL PERSPECTIVE. In recent years there has been a resurgence of interest among sociologically-oriented philosophers of science in the historical sociology of science in the modern European nations and the USA—especially in the nineteenth and twentieth centuries. Some of the work done in this field seems more easily tied to the 'empirical' referents than the work we have been describing. Unhappily, the range of sociological issues which come into view in this more limited perspective are likely to be narrowed by the bracketing of civilizational and intercivilizational settings. Principal contributors to this latter field include: P. Duhem (1954), J. Ben-David (1971), L. P. Williams (1962), A. Thackray (1970), R. S. Turner (1971). A recent piece by Stanley Coben (1971) on 'The scientific establishment and the transmission of quantum mechanics to the United States, 1919–32' represents this form of work at its best. Coben's essay offers the strongest contrast to the micro-social-psychological studies which constitute a large part of the work which describes itself as the sociology of science.

[1] See especially M. Clagett and others (1961, *passim*); Haskins (1960, 1961); Sarton (1968, especially vols II–III); White (1969).

References

Agassi, Joseph (1963), *Towards an Historiography of Science*, The Hague: Mouton; *History and Theory*, Beiheft 2.

Ben-David, J. (1971), *The Scientist's Role in Society: A Comparative Study*, Englewood Cliffs, N.J.: Prentice-Hall.

Clagett, Marshall (1962), *Critical Problems in the History of Science* (Proceedings of Institute for the History of Science), Madison: University of Wisconsin Press.

Clagett, Marshall, Post, G., and Reynolds, R. (eds) (1961), *Twelfth-Century Europe and the Foundations of Modern Society*, Madison: University of Wisconsin Press.

Coben, Stanley (1971), 'The scientific establishment and the transmission of quantum mechanics to the United States, 1919–32', *American Historical Review*, 76:2, pp.442–66.

Cole, Stephen and Cole, Jonathan (1967), 'Scientific output and recognition: a study in the operation of the reward system in science', *American Sociological Review*, 32, pp.377–90.

Crombie, A. C. (ed.) (1963), *Scientific Change*, New York: Basic Books.

Crozier, R. C. (1970), *China's Cultural Legacy and Communism*, New York: Praeger University Series, Praeger Library of Chinese Affairs.

Duhem, Pierre (1954 [1906]), *The Aim and Structure of Physical Theory*, trans. P. P. Wiener, Princeton University Press.

Durkheim, E. (1915 [1912]), *The Elementary Forms of the Religious Life*, trans. J. W. Swain, London: Allen & Unwin; New York: Macmillan.

Durkheim, E. and Mauss, M. (1963), *Primitive Classification*, (1st ed. 1903) trans. and ed. with an introduction by Rodney Needham, University of Chicago Press.

Durkheim, E. and Mauss, M. (1972), 'Note on the notion of civilization', (1st ed. 1912) trans. with an introduction by B. Nelson, *Social Research*, 38:4, pp.808–12.

Haskins, C. H. (1960), *Studies in the History of Mediaeval Science*, (1st published 1924), New York: Ungar.

Haskins, C. H. (1961), *The Renaissance of the 12th Century*, New York: Meridian (originally published 1927, Cambridge, Mass: Harvard University Press).

Holton, G. (ed.) (1965), *Science and Culture*, Boston: Beacon Press.

Horton, Robin (1971), 'African traditional thought and Western science' in Wilson, Bryan R. (ed.), *Rationality*, New York: Harper Torchbooks, pp.131–71.

Maine, Sir Henry Sumner (1861 [1963]), *Ancient Law*, Boston: Beacon Press.

Maine, Sir Henry Sumner (1871 [1876]), *Village Communities in the East and West*... To which are added other addresses and essays. 3rd enlarged ed., New York: Murray (contains the Rede Lecture of 1875 and other addresses).

Mauss, M. (1963, 1972), see Durkheim, E. and Mauss, M.

Merton, Robert K. (1970 [1938]), *Science, Technology and Society in*

Seventeenth Century England, new introduction by the author. Harper Torchbooks and Howard Fertig.

Merton, Robert K. (1972), 'Insiders and outsiders: a chapter in the sociology of knowledge', *American Journal of Sociology*, 78, 1, 9–47.

Needham, J. (1954), *Science and Civilization in China* (4 vols in 7 parts), New York: Cambridge University Press.

Needham, J. (1963), 'Poverties and triumphs of the Chinese scientific Traditions' in Crombie, A. C. (ed.), *Scientific Change*, New York: Basic Books.

Needham, J. (1969), *The Grand Titration: Science and Society in East and West*, University of Toronto Press.

Nelson, B. (1968a), 'The early modern revolution in science and philosophy: Fictionalism, Probabilism, Fideism, and Catholic "Prophetism"', *Boston Studies in the Philosophy of Science*, 3, Cohen, R. S. and Wartofsky, M. (eds), pp.1–40, Dordrecht: Reidel.

Nelson, B. (1968b), 'Scholastic rationales of "conscience", early modern crises of credibility and the scientific-technocultural revolutions of the 17th and 20th centuries', *Journal of the Society for the Scientific Study of Religion*, VII:2, Fall, pp.157–77.

Nelson, B. (1972), 'On R. K. Merton's *Science, Technology and Society in Seventeenth Century England*', a review essay, *American Journal of Sociology*, 78:1, pp.223–31.

Nelson, B. (1973), 'Civilizational complexes and intercivilizational encounters', *Sociological Analysis*, 34, 2, pp.79–105.

Sarton, George (1968 [1927–48]), *Introduction to the History of Science*, 3 vols, Baltimore: Williams and Wilkins.

Thackray, Arnold W. (1970), *Atoms and Powers. An Essay on Newtonian Matter-Theory and the Development of Chemistry*, Cambridge: Harvard University Press.

Turner, R. Steven (1971), 'The growth of professorial research in Prussia, 1818 to 1848—causes and context', *Historical Studies in the Physical Sciences*, 3, pp.137–82.

Weber, Max (1946 [1915]), 'Religious rejections of the world and their directions', trans. in Gerth, H. H. and Mills, C. W. (eds), *From Max Weber: Essays in Sociology*, pp.323–59, New York: Oxford University Press.

Weber, Max (1958), *The Protestant Ethic and the Spirit of Capitalism*, trans. by Talcott Parsons, foreword by R. H. Tawney, New York: Scribner's.

White, Lynn, Jr. (1969), *Machina ex Deo: Essays in the Dynamism of Western Culture*, Cambridge, Mass.: M.I.T. Press.

Williams, L. P. (1962), 'The physical sciences in the first half of the nineteenth century, problems and sources', *History of Science*, I, pp.1–15.

2

The sciences: towards a theory*

Norbert Elias

If one were asked to predict in which respects future theories of scientific knowledge are likely to differ from contemporary philosophical theories, or more soberly, in which respect the latter are most deficient, one could draw up a short list of such deficiencies with great ease. For the philosophy of sciences perpetuates to this day a tradition of thinking which has its roots in an earlier stage in the development of scientific knowledge. The representatives of its mainstream still proceed as if the discovery of general laws of the Newtonian type were the only aim and the main legitimization of a science. They almost completely fail to take note of the fact that since the seventeenth century many sciences, including physics itself, have discoveries to their credit which can no longer be fitted into the general conceptual mould of a Newtonian law. Examples include the discovery of the structure of large molecules, of viruses, and bacilli, the sequential order and some of the mechanisms of biological evolution, and some aspects of the development and structure of societies. They totally neglect the whole complex of problems arising from the developmental changes of sciences since the seventeenth century and especially those posed by their growing differentiation, by the emergence of an increasing number of scientific specialisms. The fact of increasing scientific differentiation and specialization itself is obvious enough. The science-theoretical problems it implies are totally disregarded. Physics has

* I am grateful to Richard Whitley, University of Manchester, and to Brian S. Martin, University of Leicester, who have read the manuscript and have made a number of helpful suggestions for its improvement.

become differentiated into a vast number of physical and chemical specialisms, among them those concerned with the development of stars, galaxies and the physical universe as a whole, and those devoted to the study of subatomic events, all held together—more or less firmly—by a set of central theories. To do them justice, one has now to speak of the physical sciences rather than of physics in the singular. Biology, too, has developed into a growing network of specialized branches ranging from those concerned with problems of evolution or with animal growth and development to those engaged in the growth and structure of cells. The biological sciences, during the nineteenth and twentieth centuries, have contributed hardly less than the physical sciences towards the solution of previously unsolvable and often unknown problems. Their representatives have worked out types of theories and concepts of their own in correspondence with the specificity of the level of integration with which they are concerned and, therefore, without counterpart in the physical sciences. The theory of evolution itself, concepts like sexual differentiation, heredity, embryo, birth, larval stage, maturity or death are examples. They are symbols which men have developed and continue to develop further as tools of inquiries into the distinguishing character of the biological level of integration and as representations of their findings. The social sciences on their part have developed from the relatively unspecialized form in which they emerged late in the eighteenth century from the prescientific matrix of philosophy into a set of divergent specialisms. Their makers and representatives, too, have worked out types of theories and concepts of their own corresponding to the specificity of yet another level of integration. Examples are Marx's theory of the development of societies, Keynes's theory of trade-cycles and Freud's theory of the development and structure of the human personality, concepts such as industrialization, monopoly of physical power, trade cycle, repression or socio-dynamics. But, in their case, quite apart from the intrinsic complexity of the social level, men have to face themselves. The demythologization of the symbolic representations at this level, therefore, is particularly difficult. An integrating central theory of society indispensable for any systematic cross-fertilization and co-operation between different social sciences is still lacking. At the present stage, the social sciences are wholly fragmented.

Thus, the development of sciences, as one can observe it, has led to an increasing diversification and specialization of sciences on at least three main levels: on the physical, the biological and the social level.

Philosophers of science, in the main, take little notice of this development. The question of the reason for this differentiation of sciences and above all for the fact that biological and social sciences differ in some respects very markedly from the physical sciences, cannot be irrelevant for a theory of sciences. But it is evidently irrelevant for philosophers; it is not a subject of the philosophy of science. Their representatives are still concerned with the properties of science in the singular. They continue the tradition of the seventeenth and eighteenth centuries when physics appeared as the supreme example of men's capacity for making discoveries about the universe helped along by their own ability to reason and to observe. They still conceptualize this ability in the manner of their ancestors as a kind of internalized godhead, a personified abstraction, as reason; and they continue to seek in vain an answer to the problem how or whether it is possible for that which they assume to reside 'within' themselves, for reason, to gain true knowledge of that which is 'without'. They continue in the same way to regard laws, and particularly the laws of nature, as the principal and perhaps even as the only possible symbolic representation of regularities or connections between observable events. Although in many areas of scientific research men have found it more helpful for the solution of problems to work out symbolic representations of other types of connections, philosophers of science persist in ignoring these developments. They are probably not aware how closely the conceptualization of physical regularities in the form of laws was initially connected with the belief that the laws of nature were the laws of god, symbols of the order which he had imparted to his creation. What was new in the eyes of the early representatives of a scientific mode of discovery was the experience that one need not rely on authoritative books or on other forms of revelation if one wanted to discover the laws that god has given to his creation, the eternal laws of nature. The high value attached to the discovery of general laws, in other words, was symptomatic of a value scale which had found an earlier expression in theology, which was taken over as seemingly self-evident by classical physics and philosophy and which persists to this day, among others, in philosophy generally and particularly in philosophy of science. According to this hierarchy of values, the discovery of something eternal and unchanging behind the observable changes appears as the highest goal of men's quest for knowledge. It finds its symbolic representation in sets of personified abstractions, established as idols in the philosophers' pantheon, all of which are symbols of eternity and many of which are representations of a set of eternal laws—abstractions such as reason and nature or as essences and existence.

The philosophers' claim to examine the assumptions underlying men's thinking and men's knowledge evidently does not extend to their own thought and knowledge. They have not examined in the past and do not examine today the rationale of this value scale. By and large, it has been taken for granted that the discovery of that which is timeless and immutable, whether it has the character of immortal gods or of natural laws is the highest reward which men can gain in their search for knowledge. The philosophers' ideal science in other words is cast in an unchanging mould. They have closed their mind to the problems of change in sciences as elsewhere. Eternal laws of nature, of reason or of logic or alternatively eternal existence, essence, meaning, everyday life or common sense, in short, unchanging universals and lawlike axioms or definitions are the cornerstones on which philosophy rests. However much philosophers may differ among themselves, this type of abstraction, and the style of thinking which goes with it, is their common trademark.

Unthinkingly, they treat it as the only possible type of abstraction, as abstraction *per se*. In actual fact, some physical sciences as well as some biological and social sciences, as one shall see, have worked already for some time with different types of abstractions. But philosophers pay no heed to observable data, to sciences as they are. They are nostalgically concerned with science as it was and science as it should be. Laws and lawlike abstractions dominated classical physics. Philosophical metaphysics and classical physics bear the imprint of the same stage in the development of knowledge. Some physical sciences and indeed, some other sciences as well, have moved beyond that stage. But philosophy, particularly philosophy of knowledge and of science, has become frozen in its tracks. Otherwise it would be easier to recognize as a fallacy the notion that all sciences in order to lay claim to the status of a science must follow the models of archetypal physics.

To some extent this assumption is probably connected with what one might call the *atomistic fallacy*. According to it, the properties of composite units can always be explained in terms of the properties of the smallest component parts. One expects to be able to determine and perhaps to explain the properties of the former if one studies and above all if one measures, the properties of the latter in isolation. Physics is concerned with the units of which all other observable units, including those explored by the biological and social sciences consist. The simplest and smallest units into which one can divide all the manifold composite units of the universe including organisms and the societies which they form, appear as the basic units, as the 'building stones' of the universe. On this assumption, physics which explores the properties of the 'building stones' of the world appears as the fundamental science on which all others are based. The study of composite units without atomistic reduction, their study as a 'whole' appears in that case as metaphysics. According to an old tradition which is still very much alive the physics-metaphysics polarity offers the only alternative between which men can choose. It is a fallacy which dies hard.

If a theory of sciences were simply based on systematic scientific inquiries into sciences, it would be easy enough to show what is wrong with the atomistic fallacy or with the related notion that all scientific abstractions are lawlike abstractions and that all scientific theories are lawlike theories. A growing number of sciences do fruitful inquiries on different lines. Their practitioners have learned to frame their problems on the basis of different expectations. They have worked out theoretical models of a kind that differ markedly from the classical archetype of mechanistic laws and theories. But they often have a bad conscience because their training and the dominant consensus in their societies has implanted in them the conviction that the work of physicists is the archetype of scientific work generally, the ideal model which all work that deserves to be called scientific ought to follow. They are disorientated because the demands of their work and the demands of powerful high status groups of scientists, administrators and philosophers are at variance.

Let me give an example. It is taken from a science-theoretical essay on embryology. But it shows how much the development of science-theoretical research is held up by the fact that arguments revolve again and again around the physical-metaphysical axis. It also shows how much one could gain from comparative studies of different sciences (Oppenheim, 1967):

> The problem of modern embryology . . . was crudely summed up as the problem of determining the degree to which particular material answers to Roux' description of *differentiatio sui or differentiatio ex alio* at any one moment. Another trouble has been, historically speaking, our constant opposition of the metaphysical to the physical: there may be a biological level too, at which one might work without retreating to the camp of the

> spiritualists and vitalists, and this is where our imagination has been and still is at its weakest. Roux saw the dilemma, as have so many others:
> 'The all too simple mechanical and the metaphysical conception are the Scylla and the Charybdis. To steer one's boat between them is difficult. Until now few have been successful in doing that.'

One phrase in this quotation indicates briefly how one can open the trap in which people are caught who approach these problems in terms of the physics-metaphysics polarity. That is the phrase: 'there may be a biological level'. The hesitation with which that is said shows that one has still to go a long way before it is clear what that means: 'the biological level' and then again 'the social level'. It will be interesting to see how much time has to pass before the meaning of these terms and the different levels of integration they represent become 'obvious' and 'common sense'. The difficulty is not due to any lack of evidence. It is the integration of the evidence which is blocked. The conceptualization at the science-theoretical level has not kept in step with the development of scientific knowledge itself. In a variety of sciences it has been shown convincingly and has been accepted by the consensus of their practitioners that there are composite units whose properties cannot be explained alone in terms of the properties of the component parts, considered as independent units. They also require explanation in terms of the configuration of the constituent parts, of the way they are bonded to each other. The configuration itself, the manner of integration of component parts, in other words their functional interdependence, has an explanatory function. That is the crucial point.

The actual terms used in different sciences concerned with the exploration of composite units whose properties depend—more or less—on the configuration of the component parts, are often very different. Conformation, shape, form, organization, integration, figuration are some of them. What can be called 'functional interdependence' of the component parts in a composite unit when seen from the side of the component parts themselves, may be called 'structure' when seen from the side of the composite unit. But it means exactly the same. As 'structure', 'figuration', 'functional interdependence', 'integration', 'mode of bonding', 'organization' or however one may call it in specific cases are all symbolic representations of the same type of determinants of the characteristics of a composite unit, their discovery has explanatory functions. It has become in a growing number of sciences the central aim of research; in these cases, it has, thus, taken the place often attributed in philosophical theories of science exclusively to the discovery of laws. Nor has this change been confined to the discovery of structures at a given moment of time. Particularly in the case of sciences concerned with a nexus that develops progressively or regressively, with units that grow and decline, their practitioners have increasingly tried to work out models of the structure of change. They have attempted to represent symbolically the way in which, and the reasons why, the momentum of any particular phase in the growth of a developing structure, the internal as well as the external momentum, leads to its transformation into another which may in turn give way to yet another stage in its continuous transformation. In short, they have tried to work out process theories. In both cases, in that of static structure theories and in that of process theories, the difference to the traditional lawlike theories is striking. Although conceptualization has hardly begun, they are themselves examples of a growth process; they are indicative of the change from lawlike to process theories of sciences.

Lawlike theories represent men's aim at penetrating behind that which happens in time and space to something beyond them that is unchanging and timeless. Like human beings and landscapes painted on a two-dimensional canvas, they are projections of a four-dimensional nexus into a two-dimensional plane. They are symbolic representations of universal and eternal regularities of connections which are expected to repeat themselves regardless of any particular position in time and space. In many cases, time and space themselves are symbolically reduced to quantities included together with other named quantities in mathematical formulas.

Models of structure, stationary or dynamic, are abstractions of a different type. They include the spatial dimensions and, in the case of models of dynamic structures, of process models, also the time dimension. They are symbolic representations of three- or four-dimensional configurations or, in other words, synchronic and diachronic stereomodels. The rise of their discovery to the position of the central aim of a number of sciences and the corresponding relegation of the discovery of timeless laws and of the use of unchanging universals to a subordinate and auxiliary position in these sciences has far-reaching consequences for the theory of sciences, of which more will have to be said later. One can best illustrate at least some of them with the help of specific examples from sciences where shifts of emphasis from lawlike to structure and process models have actually occurred.

Sciences can be arranged in accordance with the degree of integration or of structuredness of the units which they explore. A comparative study of different sciences along these lines is instructive. At the one end of the scale are to be found sciences concerned with composite units, nexus, fields or whatever one likes to call them whose component parts are so loosely connected with each other that their functional interdependence approaches the zero point. Processes of integration and disintegration of such composite units, therefore, are reversible. The constituent parts can be separated and reassembled in the same manner. The structuredness of such composite units is minimal. The properties characteristic of their component parts even if they are studied in isolation outweigh those of their configuration, of the pattern of their interdependence as determinant of the properties of the composite unit. Nearer to the other pole of the scale, are composite units such as organisms which grow and decay, which have not only a static but a dynamic structure. In their case, processes of integration and disintegration, in the last resort, are irreversible. It is not only that the constituent parts which belong to the same level of integration as the composite unit itself cannot be reintegrated with each other after the disintegration of the latter. The structuredness of composite units of this type or, in other words, the interdependence of the constituent parts is so high that the independence of these parts, though not entirely lacking, is very small. Hence, the properties of their configuration, of their functional interdependence, or, in other words, of the structure of the composite unit outweigh by far those of their component parts considered in isolation as determinants of the properties of the composite unit. A somewhat different picture emerges if one examines integration and disintegration processes of human societies. But that can wait. It need not be done here. All that is needed at the moment is a preliminary picture of the configuration of sciences that emerges if they are arranged in accordance with the degree of structuredness, with the balance between relative independence and relative interdependence of the constituent parts, characteristic of the composite units which they set out to explore.

An example of a scientific specialism devoted to the exploration of units with a very low degree of structuredness is the chemical study of gases. For a gas is a composite unit whose component parts are comparatively loosely connected with each other. Except in the case of collisions, the component molecules move as nearly as possible independently of each other and whatever interdependence there is diminishes as the pressure of the gas is reduced. The more that is the case, the looser becomes the connection between the molecular movements, the greater becomes the approximation of the characteristics of a gas to the ideal of atomism. The more appropriate too becomes in that case the conceptualization of regularities of the behaviour of a gas in the form of a general law. Boyle's Law[1] about the behaviour of gases, one of the best known classical chemical laws, illustrates the point. Observable gases do not strictly behave in accordance with this and other 'gas laws'. Their behaviour approximates more closely to the 'law' as the pressure of the gas is reduced. The more that is the case the more appropriate it is to regard a gas as an assembly of independently moving particles and the better does the behaviour of a gas agree with Boyle's Law. The law in other words is set for what is technically called a 'perfect' or an 'ideal' gas. The 'perfect' gas, in other words, is the gas which most closely agrees with the ideal of atomism. Like many other symbolic representations of regularities of connections between observable events in the form of a natural law, Boyle's Law is an archetypal abstraction, an ideal type construct. The fit of a lawlike, an ideal type representation of the connection of part-events of a nexus, one might say, increases or diminishes in inverse ratio to the degree of integration of the nexus. The higher its integration or, in other words, the greater the functional interdependence of its parts, the less well fitting are lawlike ideal type models, the more appropriate are other types of abstractions such as configurational or structure models, either synchronic or diachronic, as symbolic representations of connections and their regularities.

It may be useful to add that lawlike representations can have other than purely cognitive functions. They often simply have didactic functions. They help in the preliminary exposition of a problem. That is the task of the lawlike formula concerning the fit of lawlike representations. It draws attention to the task of the examples from a number of sciences which will be used here and of which that of Boyle's Law is the first.

One further general observation may be useful. It concerns the concept of order as a characteristic of a nexus of events which sciences

[1] 'At a constant temperature, the volume of a given quantity of any gas is inversely proportional to the pressure upon the gas.'

explore, e.g. the order of nature or the order of society. According to my knowledge, whenever this problem has been discussed, it has been assumed, implicitly or not, that 'order' is the opposite of 'disorder' and that whenever any 'order' is scientifically explored, disorder and chaos can be expected as its antithesis.

It is usually not made sufficiently clear that the term 'order' can be used in two distinctively different senses. In the first sense, it expresses what people experience as order in contrast to what they experience as disorder. That is especially clear in men's experience of society where the term 'order' is closely linked to the term law. What is experienced as order or disorder in that sense depends on the I—or We—perspective of specific persons or groups. Many people, though not all, may regard, in that sense, wars and revolutions as social disorders; they may, in that sense, regard an earthquake, the flooding of a river valley or the outbreak of a volcano as a disorder of nature. This use of the term order as the antithesis of disorder, in other words, is an expression of a specific subject-centred value scale. Seen independently of any particular group and in that sense 'objectively' neither the term order of nature, nor the order of society has its counterpart in any disorder or chaos. In terms of a scientific, an object-orientated inquiry neither the dissolution of a galaxy, nor the destructive eruption of a volcano is a disorder of nature. They form part of exactly the same 'order' of nature as the course of the planets around the sun or a rich field of bearded wheat swaying in a breeze. The term 'chaos' applied to nature is meaningless. So is therefore the philosophical question how to account for the 'order of nature', the fact that natural events follow 'laws'. The wasteland of the moon, the birth of a monster, pestilence, flood and hurricanes, senseless, chaotic and contrary to what men may regard as order, are as much part of the order of nature as the eternal laws of heavens, admired by Kant and likened by him to the moral laws in our breast. To some extent one can say the same if the term 'order' is applied to society. In terms of a scientific, an object-orientated inquiry, neither war, nor revolution, neither murder nor concentration camp and genocide is a disorder of society. They form part of the same order as the division of labour in a hospital or a game of football or chess. Only when seen from the I—or We—perspective of specific groups can 'social order' appear as an antithesis to 'social disorder' and 'chaos' or co-operation as antithesis to conflict. If one is concerned with the explanation of nature or society as an order of a specific type, as one level of the multi-levelled universe, one is confronted with a nexus of events that includes anything which men according to their subject-centred value scheme may classify as disorder or chaos. It would be better to standardize two different terms for 'order', namely 'order' in a factual sense and 'order' as an expression of what specific groups or persons experience and evaluate as such in relation to themselves. Using the conventional terms, one might speak of 'subjective order' and 'objective order'. Only if used in the former sense is order the antithesis to disorder. In the second it-centred sense, 'order' cannot be used as antonym to 'disorder'. In terms of the explanation potential, there is no chaos, apart from the fact that men themselves experience and evaluate certain conditions of nature and society as chaos and disorder. Many pseudo-problems could be discarded if this distinction was clear.

The term 'order' used in a factual sense, means nothing more than that every observable event is part of a nexus of events. It means that the connections of events with each other can be explored, discovered and explained by men and that the structure and regularities of connections, the position and function of particular events within a nexus of events, can be determined and symbolically represented by concepts, models and theories whose fit is open to tests and improvements. This concept of 'objective order' is based on the present fund of human knowledge. According to that fund of knowledge, no event occurs totally and absolutely unconnected with others. No event, in other words, is unexplainable. Sciences are based on the experience of previous generations and accordingly on the expectation that whenever at a given stage of knowledge the connections of events with the already known nexus of events are unknown, such connections exist and can be discovered by men. That and nothing else is meant by 'order' in the object-centred sense of the word. If it is possible to find evidence for the occurrence of totally unconnected events, one will have to reconsider the matter. That would demonstrate the occurrence of disorder and chaos in the object-centred sense of the words. As long as that is not the case, it is useful to keep in mind that the development of the natural and to a lesser extent that of the social universe, under prevailing conditions, is indifferent to the distinction made subjectively, i.e. from the first-person-perspective of particular human beings, between order and disorder.

At first, it may be hard to face the fact that the course of nature and, under so far existing conditions, also the development of human society, of the commonwealth of men, are neither ordered nor disordered in the subject-centred sense of the words, neither inherently benevolent nor malevolent, but merely blind and indifferent to the fate and feelings of men. Much of the present conceptual equipment including metaphors like 'the laws of nature' still bears the imprint of men's dream of a world which is basically ordered for their best, either by superhuman beings or by personified abstractions such as benevolent nature. Law and order in nature, thus, appears as the prototype of law and order in society. In actual fact, the social ideal is primary; lawful and well-ordered nature is its projection. Both are ideals disguised as facts. The scientific discovery of laws has been frequently misunderstood as the supreme ability of scientists to reveal the preordained orderliness of nature, of reason or of society.

Thus, the reference to a particular law may help to clear away, at the same time, some of the misapprehensions surrounding the functions of sciences and the nature of general laws.

Boyle's Law can serve as a simple example in both senses. A jarful of gas is a relatively loose composite unit. It would rank relatively low on a scale on which units explored by sciences are arranged according to the degree of interdependence of their component parts or, in other words, to the degree of their structuredness. In that respect, as one saw, it approximates to the atomistic ideal. The immediate component parts of a gas, the molecular movements, have a relatively high degree of independence in relation to each other. Compared with solids or liquids, gases, in the language of involvement, represent a condition of greater disorder or chaos. The more it approaches that condition, the better its properties agree with the prescriptions of Boyle's Law; the more it assumes the characteristics of a perfect or ideal gas. One might say that in this, as in a number of other cases, the greater the disorder the better the fit of the law.

If one aims at a theory not of science, but of sciences, if, in other words, high theoretical significance is attributed to the fact that sciences have diversified, the problem of the reasons for the emergence and development of different groups of sciences and of the relationship between them moves into the centre of one's attention. It should not be forgotten that men like Comte and Spencer in the first half of the nineteenth century already raised such problems. Comte, as far as I know, was the first to formulate them clearly and to attempt an answer. He tried to explain the difference and the relationship between the main groups of sciences in terms of the growing complexity of their field of research. It was not a bad guess. In fact, one can say that this explanation still stands. But, the advances which sciences have made since that time make it possible to go further. We are now able to link the scale of increasing complexity to that of increasing structuredness. What has been said before about the relative lack of structure of gases has prepared the way.

The difference becomes very much apparent if one takes as example one or two of the problems which one has encountered in studying the structure of molecules. Even relatively small and simple molecules present problems of integration, of the manner in which the immediate constituent parts of molecules, their atoms are held together. The first step on this road was the solution of the problem: what elements, what kind of atoms form together the molecules of a particular substance and in what proportions, in which quantities? This stage has found its representation in the quantitative, the so-called 'empirical' formulae of particular substances, such as H_2O for water, indicating that a water molecule consists of two hydrogen and one oxygen atoms. At the next stage, one discovered that it was not enough to know which kind of atoms and how many of them form a particular molecule. This knowledge alone, it was found, was not enough to explain the properties of the molecule. One discovered that in order to solve this problem one had to determine the actual position and the interdependence of the constituent atoms of a molecule. It was found necessary, in other words, to develop the purely quantitative law-like formula which abstracted from all references to spatial dimensions, which represented the atoms of a molecule as if they were all situated in the same plane, into a three-dimensional representation of a molecule indicating exactly how the constituent atoms were placed and bonded in relation to each other. In terms of a theory of sciences, this transition from lawlike and purely additive to three-dimensional representations of molecules, to models of their structure has the crucial significance of a signpost. It is one of the earliest indications that the physical sciences themselves, even when dealing with atoms in practice though not necessarily in their self-understanding, are developing beyond atomism towards what has been called 'structuralism'—a term now much abused. In fact, it is only if one pays attention to the comparative development of sciences, that the meaning of the term 'structure' and the reason why this concept, together with that of process, has come to take in many sciences the place formerly held by the concept of 'law', can become clear.

One can illustrate, at this level, the science-theoretical relevance of the transition from lawlike symbolic representations which abstract from spatial interdependencies to structure models which include them, by reconstructing briefly one of the problems that induced representatives of chemistry and physics to move in that direction—the problem of isomers. They became aware of the fact that substances whose molecules consist of exactly the same kind of atoms in exactly the same quantities can have different properties which means that they are to all intents and purposes different substances. The observation contradicted one of the basic assumptions of an atomistic belief system—the assumption that the same component parts in the same proportions result in identical composite units, in units with the same properties.

Take, for instance, a monosaccharide called hexose, a sugar with six carbon atoms whose quantitative or 'empirical' formula is $C_6H_{12}O_6$; it consists of six carbon atoms, twelve hydrogen atoms and six oxygen atoms. This is the formula for gluctose, the sugar present in the blood. But fifteen other sugars, among them fructose and galactose, have the same quantitative formula; they have the same component parts in the same quantities although they differ with regard to their solubility, their sweetness, their organic functions and in other respects. They are recognizably different substances. The problem is to explain how it is possible that different substances can have the same atomic composition. The solution, as has been found, lies in the different structure of the molecules. The manner in which the atoms are bonded to each other, their spatial configuration is different in each of the fifteen substances. Although they can be represented by the same quantitative formula, their representation in the form of a three-dimensional model showing the spatial arrangements of the atoms is different.

The solution of the problem of isomers is by now well established as a fairly elementary piece of physico-chemical knowledge. Three-dimensional models of their structure have been worked out for larger and increasingly complex molecules with a high degree of fitness to the relevant evidence. The explanatory functions of such models have become particularly well known in connection with the discovery of the structure and composition of some very large molecules of the deoxyribose type of nucleic acid or, in short, of the DNA molecule. The context problem is that of the high intergenerational stability and constancy of all kinds of living things in spite of their great diversity and variability. The basic structures which make it possible that a new organism forms itself in accordance with the models of the parent organisms appear to be the same in all living things. For some of the evidence for these structures comes from some of the simplest living things at present known, such as bacteriophages. The more specific problem was: how is it possible that large molecules of the same chemical composition which one can discover in the chromosomes of all the higher organisms and which, as one already knew, contain the 'substance' of heredity, can store the enormous quantity of information required for the controlled process, for the building up and the organized growth of a living thing, and in fact for the great variety and variability of living things. An answer to this specific problem which fitted most of the existing observational data was worked out as the culmination and synthesis of a vast amount of work done by many people (I think my 'culmination and synthesis' thesis fits the facts here better than Kuhn's thesis of a revolution), by J. D. Watson and F. H. C. Crick who produced in 1953 a model of the DNA molecule which fitted both the observational data and the functional requirements. The model has become known as the double helix because the two sugar-phosphate chains of the DNA model held together by hydrogen bonds between these bases are twisted around each other like a spiral staircase whose base to base attachments represent the series of steps. While it was indispensable to know what kind and what groups of atoms formed such a molecule, again, the crucial question was their spatial arrangement, the way they were bonded to each other, their configuration or, in other words, the structure of the molecule. For the atomic composition alone could not provide the answer to the problem. One had to know their specific configuration in order to explain the function of such molecules as intergenerational coding devices for a vast number of different organisms. The model of the configuration, of the structure of the DNA molecule worked out by Watson and Crick, appears to satisfy this requirement because the possibilities for variations in the sequence of the bases is practically limitless.

The solution of the problem of isomers and that of the DNA molecule are only two examples of the development of knowledge about the structure of molecules and of the related knowledge of the connection between structure and function. It is a development which is likely to continue. The culminating innovatory synthesis representing the solution of a critical problem in the pathway of a group of sciences opens the way for the work on a new generation of problems which prior to the solution of the antecedent generation could not be attacked and often enough not even clearly perceived and set.

In its home sciences the present knowledge of the structure of molecules commands a very wide measure of acceptance. But, the impact which this development, and a number of parallel developments in other sciences, have made on the theory of sciences is negligible. The science-theoretical innovations implied in these developments have not been brought to light. In fact even those men whose scientific imagination, whose capacity for intellectual synthesis of a vast array of details, contributed most to the innovatory solution of critical problems in their special scientific field, quite often fail to see the science-theoretical innovations inherent in their own work. In terms of a theory of sciences, they tend to interpret their own innovatory achievements in conventional terms. In this as in other cases the recognition of the innovatory science-theoretical significance of innovations in specific sciences was, and still is, apparently blocked by a whole complex of factors. Among them questions of status and prestige differentials between different sciences play no less a part than the wide ramifications of the atomistic belief system which underlies much of the present thinking about the nature of sciences, not only in philosophy, but also in the physical, biological and social sciences and in society at large. The innovatory breakthrough, when it comes, may well be regarded as 'revolutionary' in that sense of the word which most closely resembles its non-figurative meaning, namely explosive because a development long overdue is held up and blocked by an existing power structure. There is abundant evidence for a science-theoretical synthesis, for a scientific theory of sciences which sooner or later will take the place of philosophical theories of science. But even those people who go some steps in that direction usually still do so very hesitantly, perhaps because they are afraid that any development beyond the traditional atomistic and mechanistic belief system can only lead into metaphysical vitalism and transcendentalism.

An example of this reticence may help to illuminate the point. In a recent paper 'Atomism, structure and form', Lancelot L. Whyte (1965) summed up some of these trends in the development of the natural sciences in the following manner:

> It seems that the method of treating the properties of ultimate small units as fundamental only works well when a rather small number of variables—say less than five or ten—is adequate for some purpose, as in simple molecules and homogenous fluids. When it comes to complex partly ordered systems rich in structure and form, this classical method is very clumsy if we are interested in ultra-structure

> and its changes. Nature seems to be saying with all possible emphasis: The Laws of complex systems are not written in terms of quantitative properties of localized constituents but in terms of... ?

The answer to this question, in the light of the examples that have been given here, is quite straightforward. It is no longer a question of finding laws. 'Complex systems', as they are called in this quotation, are units which, on an ascending scale, are determined not only by their composition, but increasingly by the configuration of their constituent parts or, in other words, by their structure. I cannot show here that this is increasingly the case as one moves from large molecules to more and more highly integrated units with more and more levels of integration. For the time being, the science-theoretical lesson to be drawn from scientific inquiries into the structure of large molecules must be enough. If one examines the science-theoretical implications provided by the discovery of the structure of very large molecules with a specific biological function, such as that of the DNA molecule, the conclusions are these. Its structure as well as its functions have been explained with the help of (a) an analysis of its component parts, (b) a synthesis in the form of a testable three-dimensional model showing, in the light of the available evidence, the configuration of the component parts, (c) an examination of the way in which composition and configuration, i.e. the structure of the molecule accounts for its function. Traditional theories of science tended to concentrate attention on analysis alone. Synthesis and integration of theoretical constructs as the counterpart of studies of 'wholes', 'systems', patterns, forms or however one has called them have long been suspect as symptoms of extra-scientific speculation, of a metaphysical frame of mind and, more often than not, the suspicion was justified. The actual development of sciences, of which that of crystallography and biochemistry are only some examples, very clearly indicate that this need not be the case. Men who stand in the hardened tradition of atomism may continue to believe in the exclusivity of analysis of the component parts of a composite unit as the key to a scientific inquiry. It is this tradition which has surrounded inquiries into the integration of constituent parts, into the structure of a composite unit and the building up of three-dimensional or in other cases of four-dimensional models, as distinct from planar lawlike theories, with the odour of a non-scientific occupation. The procedure adopted at the level of intramolecular structures, paradoxically enough, is in that respect symptomatic. It heralds the break with the traditional atomism. It shows that analysis, i.e. the determination of the properties and proportions of the component parts of a composite unit, though indispensable, is not enough. Like the quest for lawlike connections, analysis plays increasingly the part of an auxiliary operation in relation to synthesis and the working out of structure—and process—theories, as one moves from the investigation of loosely integrated units at one and the same level to the investigation of more and more highly integrated units with more and more levels of integration.

To use once more the example of the DNA molecule, one could not have worked out a well-fitting model without a full knowledge of the chemical composition of the various bases, sugars and nucleotides forming a DNA molecule. One had to determine, for instance, the stuff composing the bases. One had to know that they contained purines, namely adenine and thymine, and pyramidines, namely guanine and cytosine, both composed of carbon and nitrogen in varying numbers. One needed a clear understanding of what the component parts actually do, both on the atomic level and on the level of small molecules; one had to know what properties they have and how they behave. But all the knowledge of the behaviour of parts which one could gain by means of a chemical analysis was by itself quite insufficient. Analysis was an indispensable auxiliary procedure in relation to the main task—the building up of an integrating model. One had to find out not only that in a DNA model many thousands and perhaps even millions of nucleotides are joined together. One had to discover above all how the component units were arranged within the long polynucleotide strands in order to explain the biological function of these giant molecules as templates for the building up of another generation of living things.

Thus, analysis stood here in the service of synthesis. One had to know the characteristics of the component parts, but one could not from this knowledge alone deduce either their configuration or the properties of the composite unit. The consequences of the fact that in the case of studies of more highly integrated units the reduction of the unit, its dissection into component parts by itself is quite insufficient as a scientific procedure—its consequences for a theory of sciences are far reaching. It means in essence that sciences concerned with the study of more highly integrated units cannot fulfil their task without the knowledge provided by the scientific specialists concerned with the study of the lower level component parts, but they cannot be reduced to the lower level sciences. The practitioners of the latter cannot on their own explore and determine the configuration of the component units, the structure of the composite unit; they cannot determine the specific properties of this unit for which the type of integration of the component parts or, in other words, the structure of the composite unit provides the explanation.

At this point, one may begin to see the answer to the question raised earlier, to the question of the reason why sciences of different types have developed in course of time. As one ascends from studies of simpler and loosely integrated to that of more and more complex and more firmly integrated units with a growing number of superimposed levels of integration, the instruments of research which one has developed in order to explore the relatively simple and loosely knit units, among them purely analytic procedures or the concept of general laws, lose the exclusiveness and primacy which they possess on their own level, and different concepts, such as integration and disintegration or structure and process, and different procedures, such as the building of integrating, configurational models, gain the ascendency. The notion, which still lingers on, that all sciences may eventually work in accordance with the models of physics or may even transform themselves into a kind of physics, is a mistake. One cannot adequately study and explain units at a higher level of integration in terms which have been worked out and which have been found reasonably adequate for the study of units representing a lower level of integration. Nor can one expect that a method of research which has been found adequate for scientific studies of units at the lower levels will automatically prove equally adequate for the study of units which represent higher levels of integration. As one can observe them, the methods of sciences are far more differentiated than the popular expression 'the scientific method' suggests. To put it at its simplest, there are considerable differences between the methods to be adopted in order to accumulate data with the expectation of discovering laws or lawlike regularities and the methods required for working out and for testing three-dimensional models, such as that of the DNA molecule, whose structure has not only to fit a vast number of measurements, but also specific biological functions. Like science in the singular, 'scientific method' is an ideal type construct. The common task of sciences is that of solving problems in a testable manner and of setting problems in such a way that they can be solved in that manner. The tendency to regard one particular type of method, of theories, of concepts or of abstractions, as 'scientific' to the exclusion of others accounts for many difficulties which one can observe in the past development of sciences. It creates analogous difficulties today.

Perhaps an example may help to illustrate the point. It may supplement the examples given before because it is taken from a much higher level of integration and can indicate how many new problems may come within our grasp if we do not proceed on the assumption that a theory of sciences can be reduced to a theory of science. If one studies the scientific development leading up to Darwin's theory of evolution it is noticeable that this, too, was a case in which the crowning innovatory theory represented not so much a revolution—the counter-forces were strong, but by no means able to block the innovatory trend for a considerable time—but the consummation and synthesis of a fairly long preceding development on both the theoretical and empirical levels. Darwin's theory was one of the first process theories which has shown a very high degree of fitness. In terms of a theory of sciences it represented, like that of Marx, the transition to a new type of theory. Structure theories, in contrast to lawlike theories, embody the spatial dimensions. They have, as one saw, the character of three-dimensional models. Process theories have the character of four-dimensional models. In contrast to lawlike theories, they do not abstract from either the spatial or the time dimension. If one looks at the seminal conflict within which the first biological process theories developed—Darwin was not the only man who at the time worked out a biological process theory, but only the most successful—one may notice that many of his opponents differed from him not simply with regard to the details of their arguments, but also with regard to the type of theory which they put forward. They persisted in working out lawlike theories. They tried to find the eternal laws of animals, or lawlike regularities which specific types of animals had in common. I have already alluded to the fact that the transition from lawlike to structure and process theories involves a reorientation of the whole manner of thinking. One can see it here. Among the main opponents of an evolutionary theory of living things were Georges Cuvier and Richard Owen. Cuvier, as comparative anatomist and palaeontologist, had a number of important scientific advances to his credit but he remained in his whole outlook fixated to the idea that the scientists' task is the discovery of that which is unchanging in all the variety of changeable living things. He, therefore, aimed at a type of abstraction indicating what all animals have in common. He tried to show by means of a once famous principle of the correlation of parts that the structural and functional aspects of anatomy can be regarded as part of a single universal law which is applicable to all animals. He also tried to formulate laws of correlation which showed what was common to specific types of animals (1795). An example is: 'All the animals with white blood who have a heart have bronchi or a clearly circumscribed respiratory organ. All those who do not have a heart, have no bronchi. Wherever heart and bronchi exist, there is a liver; wherever they lack, the liver is lacking.' Lamarck who prior to Darwin was perhaps the most prominent proponent of a biological process-theory shortly afterwards suggested abandoning the use of the blood colour as determinant of the classification of lower animals and introduced in his *Système des animaux sans vertèbres* (1801) the division of animals into vertebrates and invertebrates which has shown its greater object-adequacy by its continued use up to our own time. But he combined the quest for type criteria with the knowledge that the types are fluid, that they are phases or stages of a process. Cuvier, on the other hand, persisted in thinking of different classes or types of animals as eternally set in the same mould. He, therefore, aimed at discovering behind the manifoldness of observable data a lawlike ideal type of different classes of animals. This style of thinking was equally pronounced in the theories of his English counterpart and admirer, Richard Owen, who worked out a theoretical model by means of which he was able to reduce all the various vertebrate classes to a single archetype representing the universals of the vertebrate skeleton.

The example may help to show how similar the conceptual problems are which one encounters in the transition from lawlike to process models, even though in substance the processes which one tries to explore may be very different. The process of biological evolution is very different from the non-biological process of social development and as part of it the process of scientific development. Each has a distinct pattern of its own. But, in all these cases the seminal conflict between representatives of lawlike theories and of process theories show certain common structural features. The quest for the common characteristics of animals or of specific classes of animals, in short, for the universals of animals, was not unfounded. But, as the problem of the development of living things moved into the centre of the attention of biologists, the problem of the universal, of the ideal type animal, without losing its function entirely, became a subordinate problem compared with that of the evolution of the process of living things. A similar change can be expected in the theory of societies and of sciences. So far, the problem of the archetypal science has dominated the theory of science. So has the notion that one has found it in a kind of ideal type physics and in the implied assumption that all sciences can, will or should be reduced to the model of that ideal science.[1] By implication, advances in scientific research, of which I have given some examples, indicate why this reduction is bound to fail. But the implications of these advances for a theory of sciences have been largely neglected and obscured. Some of them have been shown here. The key to the puzzle is the discovery that the configuration of part-units as determinant of the properties of a composite unit is irreducible. In course of time this discovery will require the abandonment of the atomistic and reductionist theories of science in the singular. With the increasing diversification of sciences they have become fossilized relics of a former age. Today one requires a theory of the scale of sciences symbolically representing the scale of the universe which sciences explore. For at each level of that scale one encounters types of connections, of structures and processes not to be found at the lower levels. To some extent, therefore, different symbolic representations and different procedures are needed for the exploration of different levels of that scale.

References

Cuvier, G. (1795), 'Anatomy and kinship relations of worms', *Décades philosophiques, littéraires et politiques des ans*, III, V.

Elias, N. (1956), 'Problems of involvement and detachment', *British Journal of Sociology*, vol. 7, no. 3, pp. 226–52.

Oppenheim, Jane M. (1967), 'Analysis of development: problems, concepts and their history', *Essays in the History of Embryology and Biology*, Cambridge, Mass. and London: M.I.T. Press.

Owen, R. (1848), *Archetype and Homologies of Vertebrate Skeleton*, London: van Voorst.

Whyte, L. L. (1965), 'Atomism, structure and form' in Kepes, G. (ed.), *Structure in Art and Structure in Science*, London: Studio Vista.

[1] A brief exposé of a non-reductionist model of sciences can also be found in Elias (1956), pp. 241ff.

The Social Analysis of Scientific Development

3

On a sociological theory of scientific change

Peter Weingart

Problems of the sociology of science after Kuhn

Kuhn's influence on the sociology of science has proved to be so profound that he has all but attained the rank of Merton. It seems he is even replacing him, if one takes account of the change in the basic constellation of problems of the sociology of science which has emerged. Thus, we presently witness a 'problem shift' which leads the sociology of science from what, nearly exclusively, has become the study of the behaviour of scientists and the internal organization of science, its system of values and norms, communication and growth processes as well as the processes of social evaluation, control and stratification back to a perspective related to the sociology of knowledge, in that the cognitive processes are taken to be problematic and are seen as interacting with social factors (cf. Barnes and Dolby, 1970; King, 1971; Martins, 1972; Whitley, 1972). Such an approach, as has been pointed out, presupposes the abandoning of the positivistic epistemology according to which the growth of scientific knowledge is a cumulative and one-dimensional process and that the scientific 'method' as well as the logical criteria, from which it is derived, are of timeless validity (cf. Merton, 1957; Storer, 1966, pp.73, 100). Kuhn, more pronounced than other current philosophers of science, has challenged this view with his historical account of 'revolutions' in the development of science and the historicity of scientific theories and standards implied therein. And, moreover, he has drawn attention to the need for changes in the cognitive structures to be explained partly in terms of social factors. Thus, a new task is bestowed upon the sociology of science which, once accepted, will change its traditional mode of inquiry. Kuhn may be looked upon as the forerunner of what can be called a substantive sociology of science, which is possible and necessary: (a) if the growth of knowledge points up discontinuities and does not progress cumulatively, (b) if the standards by which decisions about the scientific character of theories or paradigms or, more generally, cognitive orientations, are made vary over time or can be interpreted differently, or at least leave open a range of uncertainty with regard to the decisions which result from these standards, (c) and, therefore, the acceptance of new cognitive orientations does not follow automatically and always in one direction, but is rather a social process which entails the achievement of consensus and the internalization through specific socialization agents, (d) if the validity of certain theories, paradigms, or standards is coupled in such a way with stratification and status in science that it gains authoritative power. Conversely, this implies that its change does not occur exclusively according to consensual, epistemological criteria, but has to be explained also in terms of 'influence' and structural 'means of power'.

Dimensions of scientific change

The four above mentioned stipulations are, of course, interconnected. Only the first two, however, are directly concerned with the 'cognitive substrate' of the sociology of science, i.e. the cognitive content of science. Before the interdependence of the 'cognitive substrate' and the institutional structure in scientific processes is studied, the 'substrate' itself has to be decoded. A closer look at Kuhn's concept shows, that with regard to the definition of paradigm, scientific demarcation criteria, conceptual systems, general values, etc. there is no agreement or clarity, and many of the controversies over the historicity of 'science' are due to the lack of conceptual precision. Neither the rather anecdotal history of science nor the normative philosophy of science have yet contributed to clarification, because they proceed either unsystematically or non-empirically. A conceptual differentiation of the 'cognitive substrate' is necessary, however, if different processes of cognitive change and their dimensions are to be identified as well as the laws they are subject to and the factors which influence them. In the following this problem will be clarified with examples from the current discussion, primarily of Kuhn's ideas. It must be understood, however, that this essay does not claim to offer a conceptualization of the elements of scientific knowledge—for which the sociology of science will still have to rely on the philosophy of science—but rather draws upon the latter in the attempt to analyse these concepts in their sociological context.

As a starting point, one may take the alleged difficulty of Kuhn's thesis that social and psychological factors determine the formulation of new paradigms. The element of arbitrariness in scientific development which seems to be implied by that, is confronted with the apparent evidence of unidirectional development according to criteria of epochal validity. Formulated differently, the question would read: how does one account for the fact that the growth of knowledge proceeds progressively and cumulatively when it is supposed to be determined by decisions of scientists which cannot be entirely deduced from an invariable system of cognitive orientations?

One of Kuhn's answers relativizes the presuppositions of this question altogether and should be mentioned first. According to this view, the unidirectionality of scientific progress is time and again re-established as an illusion in science education due to the revision of textbooks. The alternatives of the development are not in effect available to the individual, nor are they documented in the written history of the discipline (cf. Kuhn, 1962, 1963). This argument becomes somewhat plausible in view of the fact that a critical historical or philosophical reflection is not a component of scientific training, especially in the natural sciences (cf. Martins, 1972, p.53). The opening question, however, will be taken seriously for analytic reasons. In order to answer it, it must be made clear *what* is changing in science and *how*, that is, in what spans of time.

An answer to the question as to *what* changes presupposes a general remark concerning the social quality of cognitive orientation. Whether one speaks of values and norms or of paradigms, beliefs, methods, etc. in the context of the philosophy of science, in so far as they constitute science as a social activity they must be shared by a group of people and to some extent be identifiable in their behaviour or thinking. Therefore it can be said, as Kuhn in fact does, that they constitute a community or a multitude of them, however they may be structured. Thus, methodologically they have a status similar to that of social values and norms in sociological theory, except that the term 'paradigm', or 'cognitive orientation complexes' as it will be termed here, entails more than that of 'norm'. While the former may have a normative nature, as is documented in their institutional dimension, they are also subject to recall on the basis of insight and rational discourse. This ambivalence is not covered by the sociological term 'norm'.

Kuhn's answer to the problem of continuity ascribes a directive function to certain 'paramount values' which guarantee scientific progress. These values, which are similar to Merton's norms, have an important role 'at times when a choice must be made between theories' (Kuhn, 1970b, p.21). It must be noted here that the process of socialization in science also includes norms of behaviour and values which are of a more general type than paradigms, and which have their correspondence in the institutional structure of science. Without speculating about the content of the norms and values, one can say, with Kuhn's reservation, that this in a very formal way is the type of idea that must be sought (cf. Kuhn, 1970b, p.21).

Even under the presupposition of a 'transparadigmatic' system of values and norms, the contradiction remains, that on the one hand it is supposed to ensure continuity (in an apparently more general sense than paradigms themselves), but on the other hand allows for thoroughly divergent and conflicting interpretations and does not always determine behaviour unequivocally. The latter is a central theme of the sociological theory of action, and the theoretical juncture, where normative institutionalization of action (here, the scientist) loses its effectiveness and authority as the subsequent 'enunciation of prescriptive or prohibitory rôle expectations by occupants of responsible rôles' is instituted (Parsons and Shils, 1962, p.203). This will be important in a later context. Norms and values are always ambivalent to a certain degree and they are different in centrality. They alone, then, cannot guarantee the continuity, i.e. the 'integration of the scientific system'. Since here only the different 'levels' of change will be conceptually clarified and not yet the mechanisms that structure it, it may suffice to say that a general system of values and norms must, by definition, be *more permanent* than the paradigm, and that it may change *independently* of them, or at least in different intervals.

The difficulty of determining 'what' it is that changes comes into focus with Kuhn's ambiguous definition of the concept of 'paradigm', documented by Masterman's identification of twenty-one different meanings of the term in Kuhn's text. She classified these in three overlapping categories. Metaphysical paradigms correspond to a metaphysical way of looking at things, a cognitive or perception orienting principle; sociological paradigms are concrete, generally recognized scientific accomplishments; construct paradigms (or 'artefact paradigms') lastly, are textbooks and classical texts or a certain instrumentation (cf. Masterman, 1970, p.65). Regarded sociologically, paradigms are sets of 'scientific habits' which by Kuhn are not used synonymously with 'scientific theory'. They are 'pretheoretical' orientation complexes, which are 'sufficiently unprecedented to attract an enduring group of adherents away from competing modes of scientific activity and sufficiently open-ended to leave all sorts of problems for the redefined group of practitioners to solve' (Kuhn, 1962, p.10).

Unclear as the terms remain in this form, it is at least understandable that the process of change which they involve must be of a long-range nature. Traditions, ways of pursuing science, or 'habits' as 'units of choice' (Kuhn, 1972, p.2) or scientific decisions presuppose lengthy processes of change of attitude, as Kuhn consistently implies, and are primarily guaranteed by socialization. With this conceptual framework, the idea of *revolutionary* change is fixed in the same way, not in the sense of time, but rather in regard to content. One of the central questions in studying the structure of scientific change is that of the range of accepted applicability of paradigms. With respect to that, again Kuhn is ambiguous. Masterman's categorization of paradigms illuminates their different range, and Kuhn, too, speaks of 'very specialized' paradigms which are evidently differentiated from more general ones (cf. Kuhn, 1962, p.10; 1970a, p.252). They are by no means unequivocally limited to subdisciplinary specialties, as Martins assumes (Martins, 1972, p.11), and although Kuhn actually sympathizes with this limitation, he sees many exceptions to the rule.[1]

From an institutional viewpoint the central unit for him seems to be, above all in the mature sciences, the 'specialists-groups'. 'The analytic unit would be the practitioners of a given speciality, men bound together by common elements in their education and apprenticeship, aware of each other's work, and characterized by the relative fullness of their professional communication and relative unanimity of their professional judgement' (Kuhn, 1970a, p.253). For the sake of Kuhn's sociological intention it is necessary to defend the notion of these

[1] The difficulty in identifying the range of acceptance of paradigms leads Kuhn to attempt to solve this problem from the 'other end' by making the social basis of each respective paradigm a criterion of demarcation, of course getting caught in a vicious circle. Kuhn, by taking recourse to the 'community structure' which he understands to be the 'sociological basis' of his position (Kuhn, 1970a, p.252), already, however, points to the important inter-relation of cognitive orientations and the institutional structure.

'subdisciplinary' groups (one hundred scientists and 'sometimes much fewer' (*ibid.*)) against the unclear definitions of the paradigm. That leads, however, back to the question of what type of paradigmatic orientation, among the different variations of meaning of the term, establishes a normative cohesion and a social or scientific integration. Below the level of the more general scientific values and norms, specific standards of orientation with regard to content and methodology must be responsible for the unanimity of these groups.

If one disregards, for the moment, the necessary substantive determination of what the term 'paradigm' exactly means, there are good sociological reasons for defining the speciality, as Kuhn does, as the group or unit 'which produce(s) scientific knowledge' (*ibid.*), in which, therefore, the important substantive processes of problem-solving and decision-making take place. On account of the small size and the ease and frequency of communication (which presupposes not geographic proximity, but rather agreement over the contents and relevance of problems) one could here speak of the 'scientific peer group'. It is within the speciality that 'the pressure of the scientific peer group and the weight of shared scientific opinion will be maximal', if anywhere (Martins, 1972, p.19). Thus, a distinction is made from the further disciplinary and possibly the supra-disciplinary 'scientific community'. While the cognitive consensus of the 'specialty', would centre on a common *substantive* and *methodological* orientation, on a more general level, a consensus over paramount values would be constitutive.

Kuhn may have had a differentiation of this sort in mind, when in more recent writings he replaced the term 'paradigm' by that of 'disciplinary matrix' which entails the 'entire constellation of beliefs, values, techniques and so on shared by the members of a given community' (Kuhn, 1970a, p.175). But although he does enumerate the elements of this *disciplinary* orientation complex, he neglects to structure them systematically. They are: '(a) symbolic generalizations ($f=ma$), (b) metaphysical beliefs (belief in atoms, heat as a substance or as a form of motion), (c) values to be attached to theories (consistency, predictive capacity), and (d) exemplary, concrete "puzzle-solutions"' (cf. Musgrave, 1971, p.191, and Kuhn, 1970a, p.272).

A further aspect of differentiation has been brought into the discussion by Toulmin. What is assumed by supporters of the old and representatives of the new 'conceptual scheme' is a common set of 'selection rules' which are also subject to change, however, in larger intervals than theories, for whose evaluation they serve (Toulmin, 1970, p.44).

Although here, again, one is faced with the difficulty that cognitive orientations such as 'selection criteria' are not sufficiently specified—and therefore it remains unclear what type of community they constitute, Toulmin's notion of differential change of elements of cognitive orientations is convincing. It allows for the integration of views of revolutionary and evolutionary patterns of change. The short-term perspective of scientific change will have to account for all problems presented, methods and concepts developed and evidence assembled. In a long-term perspective, however, only a few of them have survived; the original process will be observed by what seems a sequence of steps or standard facts (cf. for a complex development of this point, Ravetz, 1971, part II).

It is evident that an important condition for the study of conflicts and decisions, of crisis and change is knowledge about the inner relations or interdependence of the various elements of the orientation complexes. The presentation of the various approaches has shown that this knowledge is lacking so far. Kuhn, as well as his friends and critics, all remain equally inexplicit in their conceptual frameworks. There can be no doubt, however, that the different cognitive orientation complexes are not distinct but inter-related, mutually supporting and compatible. On the purely conceptual level it can perhaps be assumed that the hierarchy of orientation complexes according to the degree of generality, i.e. range of applicability, is identical with the hierarchy with respect to their dependencies. This would mean that changes of the higher orientation complex, however generated, necessarily lead to changes in the complexes below. This assumption is only formal and heuristic. On the content level the picture is probably much more complex. Methods, instruments or even theories, which as such may serve as cognitive orientation complexes and constitute different types of communities, may be transferred across community border lines and be used either directly or as analogies. Their development may, in one case, 'fill a gap' and enhance an evolution already underway or, in another case, make the abandoning of an entire 'world view' necessary. The complexity of the inter-relations between the different strata of cognitive orientations is documented by cognitive situations in the history of science when the acceptance of a theory or observation is dependent on the simultaneous development of other theories, methods, tools in connection with which alone they can be interpreted and are plausible. Many of the rediscoveries are cases in point. It is obviously far too early to present any general laws with respect to these complex inter-relations—if any will ever be identifiable—and thus we have to content ourselves with the conviction that such inter-relations do exist. The concrete analysis of processes of scientific change will therefore have to be directed to these inter-relations in the cognitive sphere, because in some instances they may provide the crucial explanation.

A preliminary classification according to the degree of generalization justifies itself by the assumption that the degree of generalization of the individual elements of orientation simultaneously determines their differential potential for change. It is obvious that the attempted classification cannot be more precise than the terms on which it is based. Nevertheless, it can perhaps explain that most of the theoretical conflicts about the type of scientific change result from the divergent references to different levels of scientific orientation complexes. Kuhn also sees this quite clearly when he writes: 'the gist of the problem is that to answer the question "normal or revolutionary" one must first ask, "for whom" ' (Kuhn, 1970a, p.252).

On the basis of the terms discussed so far, the following hierarchy of cognitive orientations suggests itself:

(1) The most general level may be said to be represented by what Kuhn calls the paramount values, such as the belief in the fundamental order of nature which can be understood and/or the belief that the progress of science is cumulative and each isolated discovery contributes to the advancement of knowledge. They constitute the core of the 'political culture of science' and are also responsible for a kind of 'cognitive stratification', according to which 'mature and immature' disciplines are differentiated hierarchically (cf. Martins, 1972, who connects the differing evaluations to paradigms, and unnecessarily takes recourse to the functional theory of stratification, *ibid.*, n.29). Whether these latter evaluation criteria are shared over the entire system of science is at least questionable in view of the different concepts of theory in the natural and social sciences, e.g. empirical and hermeneutic, but also causal and functional concepts of theory.

(2) On the second level the metaphysical paradigms would have to be located, 'an organizing principle governing perception itself' (Kuhn, 1962, p.120), a whole 'Weltanschauung' (Masterman, 1970, p.67) or however else they are defined. With regard to their range of validity, generalizations are hardly possible, because they have a different

importance in the entire cognitive structure of science according to their content. Thus, Kuhn (1970a, p.252) gives examples of the different range of revolutions of these 'cognitive commitments':

> Copernican astronomy was a revolution for everyone; oxygen was a revolution for chemists but not for, say, mathematical astronomers. . . . For the latter group oxygen was simply another gas and its discovery was merely an increment to their knowledge; nothing essential to them as astronomers had to be changed in the discovery's assimilation.

For different reasons, however, it is plausible to assume that changes of these 'meta-paradigms' have far-reaching and revolutionary consequences. All of the definitions subsumed under the term meta-paradigm have the property in common that they structure cognition. They are pretheoretical and, at the same time, starting points for concept and theory construction. The long-term intervals of change, which Kuhn describes as revolutions, are probably based primarily on meta-paradigms and their dependent theory and conceptual systems.

(3) On the third level the 'sociological paradigms' could be located, which Kuhn (1962) defined most clearly as 'universally recognized achievement'. They, too, are pretheoretical and, according to Kuhn, comprise 'law, theory, application and instrumentation'. They act as models, from which coherent research traditions emerge, as for example, Copernican astronomy in contrast to Ptolemaic, Aristotelian dynamics, which is followed by Newtonian, or finally, 'wave optics' as opposed to 'corpuscular optics' (Kuhn, 1962, p.10). Very similar to this is Ravetz's concept of 'standardized' facts and tools which, once they are established, become ancestors of 'descendant lattices' of problems, or are extended to problems and objects outside the original field of application. The necessary quality for their being successful is their applicability to problems 'other than those associated with (their) first creation'. They serve the functions of challenges, heuristic guides to further work and means for problem solutions (cf. Ravetz, 1971, p.194 and ch. 6). Although a demarcation to the meta-paradigms is difficult, the sociological paradigms evidently entail a more specific cognitive orientation, which is given in the elements of the 'scientific achievements'. According to Kuhn, problems are clearly defined, so that a clearly directed research activity emerges even though in different specializations (cf. Kuhn, 1962, p.10). It is probably their immediate impact on concrete action, e.g. 'By following these (scientific

habits) successful problem-solving can go on: thus they may be intellectual, verbal, behavioural, mechanical, technological; any or all of these; it depends on the type of problem which is being solved' (Masterman, 1970, p.66) that causes Masterman to designate this kind of paradigm as 'sociological'. The intervals of change, which these paradigms are subject to, are still very long, as Kuhn's examples and Ravetz's notion of standardization demonstrate. Although it is difficult to distinguish them from metaparadigms, we can assume pragmatically, however, that they change comparatively faster than the metaparadigms and that their change generally has not such far-reaching consequences.

(4) On the next lower level, the 'artefact' or 'construct paradigm' may be mentioned, a term which stands for an actual textbook, or classic work, tools, actual instrumentation or an analogy. From its specification, it is evident, however, that the distinction between the 'sociological paradigm' and the 'artefact paradigm' is not a fundamental one. Masterman (1970, p.70), in reserving the term 'artefact paradigm' for the genetic aspect of the cognitive process, defining it as 'the initial practical trick-which-works-sufficiently-for-the-choice-of-it-to-embody-a-potential-insight. . . . ,' starting off a new science in a 'puzzle solving' fashion, does not distinguish it on the basis of the same analytic criteria as the other two senses of the paradigm concept—namely as orienting perception and behaviour in the cognitive process. Accepting the idea that the 'artefact paradigm' can neither provide a metaphysical world view nor can it be derived from established scientific achievements would seem to give it both functions, that of seeing an object in a certain way and going about its investigation. With due care it would be justified, then, to rank the 'artefact paradigm' on a lower level than the ones mentioned before.

(5) On the lowest level, finally, the 'conceptual schemes' could be located, the range of applicability of which is limited to specific attempts of problem-solving, and which are integrated into 'classical works', that is, become part of the current tradition of the specialty, or established as facts, only if they are successful. They change in relatively short intervals.

The consequences of their change affect only that group of scientists which relies directly on a specific conceptual scheme and works at specific problem solutions. As Toulmin has pointed out convincingly, the effective change of 'conceptual schemes' (namely that which is documented in the direction of change of scientific tradition) has to be distinguished from the actual quantity of innovation. Sociologically, this 'lowest' level of scientific processes, on which the actual search, problem solution, and decision-making processes take place, is probably the most interesting, while, for instance, changes of 'selection procedures' and the metaphysical paradigms are more subject of a sociologically informed history of science and culture.

The differentiation of various cognitive orientation complexes according to range of applicability and intervals of change is only a preliminary attempt to identify different units of the cognitive sphere which, although they are inter-related, may be subject to differential change. This enables us, given further elaboration and study, to view the process of scientific change in a much more complex manner than the cumulative conception of it would allow. It is that conception which also underlies the continuity/discontinuity paradox. This paradox does become apparent if the perspectives of short-term analysis and long-term hindsight are kept analytically apart. A citation from Feyerabend may corroborate this procedure: 'each particular episode is rational in the sense that some of its features can be explained in terms of reasons which were either accepted at the same time as its occurrence, or invented in the course of its development' (Feyerabend, 1970, p.216; cf. for the same point Ravetz, 1971, p.126). From the cognitive level we now turn to the institutional level of analysis, trying to show that both are interdependent and interacting, thus shaping the course of scientific change.

Social mechanisms and cognitive processes

(a) *The scope of analysis in the study of cognitive and social institutionalization*

In his early writings, Kuhn (1963) spoke of the role of 'dogma', which probably indicates most clearly that cognitive orientations have a social dimension. 'Preconception and resistance to innovation seem the rule rather than the exception in mature scientific development', they 'are community characteristics with deep roots in the procedures through which scientists are trained for work in their profession' (Kuhn, 1963, p.348). In the present context, 'social dimension' means that cognitive orientations have corresponding structures on the institutional level of science; they are mediated, enforced, or preserved by them. Without such a dimension, and entirely in the framework of an epistemological picture of science, processes and phenomena such as resistance to discoveries, revolutions in science, the discipline of 'normal science', etc. could not be accounted for. The task now is to determine the relation between cognitive orientation complexes and the social structure of science, by virtue of which it will be possible to study the factors and the mechanisms of scientific change. Thus, we start out with the vital presupposition that cognitive orientations and institutional structures are inter-related and cannot be analysed in isolation from one another. Scientific processes are taken to be processes of change which may be generated on either the cognitive or the social level. They will be fully comprehensible only if both levels, cognitive and social, are seen as mediated with one another. Change processes, then, do not occur simply on one or the other level. On the other hand the relation between the two levels is not one of direct correspondence but is, rather, itself problematic. The analysis which tries to clarify the inter-relation between cognitive and social factors and their interdependent change processes is, of course, the analysis of the institutionalization of science.

Aside from the limitation that we are here concerned with a preliminary conceptual clarification rather than the concrete analysis of institutionalization itself, a word about the role of such analysis is in order. It is not, as could be assumed, subject to the attack of being restricted to the system of science, thereby neglecting all important 'external' influences. The analysis of institutionalization processes of science, on the contrary, transcends the traditional conflict between materialistic and idealistic theories according to which scientific development is seen either as a mere reflex of the general social structure or as a unidirectional process induced by ahistorical standards. The institutionalization process of science, instead, represents the transfer mechanisms through which 'external' social influences are translated into cognitive developments and these, in turn, are translated into society. Science has to be regarded as an 'intervening variable', i.e. its organization, structures, cognitive orientations and mechanisms.

(b) *The analysis of scientific change in 'exchange theory'*

Traditional exchange theory (Storer, 1966; Hagstrom, 1965) is especially relevant to this question. The mechanism by which a created achievement is rewarded and these rewards act as stimuli to produce such achievements explains, in conjunction with the corresponding normative system, a particular form of social stratification within the 'scientific community'. It also explains science as a cumulative process which proceeds according to general normative criteria. Like the general functionalist theory of stratification, the exchange theory, however, has to make questionable assumptions and leaves vital questions open. Thus, for example, it has to presuppose that achievement and reward are in direct relation to one another, which in turn implies that achievement can be evaluated unambiguously. That reputation tends to gain independence, an obvious fact in science just as in society, which retroactively affects the evaluation of achievement and leads to the ascription of authority, is interpreted as functional by these authors if it is recognized at all (cf. Merton, 1968). Treated as non-problematic and immune from sociological analysis, remains the question of the genesis of the criteria on the basis of which scientific achievement is evaluated. Finally, the problem of change, as it appears in Kuhn's analysis, phenomena of discontinuous development, or scientific conflict have to be ignored by this theory; the reason being that the analytical context would switch and stratification and reputation would have to be interpreted in terms of power. Cumulative progress and the invariableness of scientific criteria of validity are, therefore, the central presuppositions of this theory.

It appears, however, that some insights of the exchange theory can be made productive if the contradictions and gaps regarding the sociological categories and theorems can be resolved. An essential presupposition of this is that the sociological categories do not remain external to the cognitive sphere, but are related to it.

(c) *Stratification, social structure and power—the relationship between cognitive orientations and institutional mechanisms*

According to Kuhn, a paradigm constitutes the specialized communication network and therefore the 'specialty' as a social group. The paradigm specific to the group is transferred through socialization, structures the cognitive process, determines the problems and legitimates their solutions (cf. Kuhn, 1963, p.349). It is the necessity of transfer by means of socialization that requires a minimum of formal institutionalization of the paradigm and its community. This fact suggests that the scientific community, in its institutional expression, can be differentiated analogously to the different levels of the orientation complexes. Thus, following the differentiation suggested above, it could be reasoned that only the lowest level of the 'conceptual schemes' integrates a group, which is constituted by immediate and enduring communication as is given with common problem orientation, application of techniques or interpretation of data. Such a group can be labelled 'research area'. Above this level 'specialties' could be located, which are built around models or a 'limited set of mutually commensurate models which purport to explicate existing "facts" and direct further investigation'; they are, in any case, more general in scope than research areas, comprising an agglomeration of them (cf. Whitley, ch. 4, below, for the distinction between research areas and specialties, cited here where also the relation of cognitive and social institutionalization is developed in detail). Above these specialties one would find the level of disciplines, as for instance physics, chemistry, or astronomy organized around a common ordering principle and comprising specialties. Finally, the most general level would comprise the whole community of natural scientists, which is constituted over 'paramount values' or the 'political culture of science' (cf. Kuhn, 1970a, p.239, for a similar differentiation which is, however, based on different criteria).

The differently defined 'communities' also exhibit different degrees of social institutionalization. It can be assumed, for example, that the small groups which work at limited problems are hardly institutionalized beyond their informal communication contacts, and are therefore very flexible. Disciplines, on the other hand, and to a somewhat lesser degree, specialties, reflect a considerably higher degree of institutionalization. Physics which, for instance, emerged as a discipline in the middle of the nineteenth century, since then has established itself at the university. According to its internal differentiation professorships are instituted, curricula and textbooks provide for the training and socialization of students, journals and professional associations are founded, and foundations have panels which are staffed with physicists who decide over the allocation of resources to physicists. These are the forms of institutionalization as they can be found more or less in each discipline. It is evident that, having grown over long periods of time, they become more inflexible and resistant to change. From this follows that resistance to innovation becomes stronger on each level of institutionalization.

A specific expression of institutionalization of science is its social stratification. According to the exchange theory and the functionalist sociology of science, the system of social stratification in science is geared to scientific achievement and based primarily on scientific 'recognition' as a specific form of 'reward'. As far as this assumption is valid, stratification must correspond directly to the differentiation of the communication networks, because the evaluation of scientific achievements can only take place within the boundaries in which it is communicable and which are relevant for the cognitive process. That means also, that the allocation of recognition is differentiated with respect to its range of applicability in the same way as the scientific communities, that is, the social structures in which alone it can become effective.

With this formal differentiation, only the scope of the problem is brought to light. The system of stratification, namely, is linked with structural elements of the system of science; a trivial fact which, however, is interpreted as functional by the functionalist sociology of science without following its consequences. The evaluation of scientific achievement always has institutional consequences, as for example, the allocation of professorships, stipends, research resources, institutes, editorships, etc., be it immediately or directly dependent on the recognized reception of achievements, the awarding of prizes or honorary degrees, or similar awards. Stratification, here as in society, means more than just a non-material evaluation. It means the allocation of facilities by means of which new facilities and resources for new achievements can be attained.

Much more so, the initial situation in the socialization process decides which kind of achievement is demanded from the newcomer and possibly which limited reward and recognition he can expect, which initial evaluation his work will be given. The fact that evaluation processes are consolidated in the institutional structure indicates the tendency towards the independence of reputation. It, in turn, retroacts, mediated by institutional factors, on the evaluation processes. Scientific achievement, then, is no longer evaluated exclusively on the basis of rational criticism, i.e. in terms of the frame in which the knowledge is produced, but is ascribed on the basis of stated authority.

While this phenomenon has been documented (cf. Merton, 1968; Cole, 1970), its theoretical importance has been ignored so far. From a sociological viewpoint, Luhmann provides some further guidelines. He sees science as a system which is based on 'Ausdifferenzierung und funktionaler Selbstverständigung des Wahrheitsmediums' (Luhmann, 1968, p.151). The 'cursory orientation to symptoms' appears as a selection criterion in place of the tested achievement of the individual scientist. Such a symptom is scientific reputation, which in turn is derived from other symptoms (Luhmann, 1968, p.155). The central function of reputation is the 'steering and clearing' of the mass of information which cannot be handled otherwise than through mechanisms which reduce the complexity and adapt it to the capacity of the system. However, the relation between the 'context of truth' and the 'context of reputation' is always problematic, as these contexts are discrepant. The orientation to reputation remains unstable because, otherwise, the complexity would be completely eliminated, that is, the system of science would freeze under a dogma and no new information and insights would be accepted that could question it. In science as a 'system of decision-making', reputation differences take the place of hierarchies which, as in the political system, are related to power as a medium of communication (cf. Luhmann, 1968, p.159). The formation of reputation is the mechanism of allocation of power specific to science. It is effective in a dual way. Institutionally, the reputation criterion serves as a relay between the general system and the allocative decisions. At the same time, on the cognitive level, reputation has the function of guaranteeing the direction of science, as it is through reputation that 'the appearance of truth is regulated' (Luhmann, 1968, p.156).

Differences of reputation or stratification in science, then, are not as the exchange theory or the functionalist sociology of science would have it, indifferent to the cognitive processes guaranteeing a continuous growth of knowledge. Rather, they are expressions of power specific to science, which structures cognitive processes: on the one hand, on the cognitive level in the form of the guiding and evaluation of perception, and analogously, on the institutional level, through the allocation of positions and resources.

In the course of the discussion over his book, Kuhn has come to similar conclusions, although he leaves out the institutional dimension when he writes that 'the responsibility for applying shared scientific values must be left to the specialist groups. It may not even be extended to all scientists . . .' (Kuhn, 1970c, p.263). He arrives at the notion of a value and norm explicating élite in science. This notion is nearly equivalent to saying that valid science is that which the respective scientists decide 'on the basis of their past experience and in conformity with their traditional values' (Kuhn, 1970c, p.263). In Kuhn's analysis of the normative, authoritative nature of the paradigm, the institutional perspective is lacking (which only recently he begins to open up, by pointing to the connection between paradigmatic consensus and community structure). It is only with this perspective that it becomes evident that valid paradigms have their institutional equivalent in a scientific élite which explicates them and whose 'power' is secured through precisely this function and the right of evaluation and allocation that are linked to it.

It has been shown analytically that cognitive processes and institutional level in science are inter-related. The problem is to account for their interdependence as well as for their relative independence. One aspect of the independence of the cognitive processes was touched upon with Luhmann's thesis that orientation to reputation remains unstable. That means, more generally, that the institutional structures which, to a certain extent, guide cognitive processes can be endangered or annihilated by the dynamics of cognitive processes, that is, unforeseen cognitive events. If the social structures of science-communities, communication, stratification, etc. are constituted by scientific values and orientation complexes, it must be possible to explain some of their changes through the dynamics of cognitive processes. Analytically, this is the basic structure of the process that Kuhn calls revolution. Conversely, and that is the other aspect, the institutional structures have their own dynamics. Institutional mechanisms such as those central to Hagstrom's analysis can no doubt determine cognitive processes independently or simply remain relatively inert to them. And finally, with some reservation with respect to dichotomies, it may be said that a stabilized institutional structure characterizes the situation of 'normal science'. Before the background of this analytical frame of reference, we can now look at different configurations of scientific change, concentrating on the question why 'autocracies surrender their power or change their evaluation' as an exemplary case (Whitley, 1972).

Cognitive and institutional determinants in the analysis of scientific change

Processes of problem change in descriptive biology, mathematics and sociology have to be differentiated from those in physics. In the formal and empirical sciences in which generally accepted criteria for the evaluation of the relevance of problems are lacking, a greater importance of 'fashions' is to be expected (cf. Hagstrom, 1965, p.181). In the case of sociology, for example, one can say that the stratification system is relatively heterogeneous; élites, whose influence extend beyond the limits of their specialty, hardly exist. Therefore, a relatively strong 'susceptibility' to 'fashions' could be expected, which results from 'extra-scientific', that is, social goals. Due to the nature of the pre-paradigmatic state of the social sciences, the goals are subject to comparatively arbitrary changes which are hardly determined at all by the institutional structure.

Hagstrom differentiates processes in which the succession and replacement of goals is a result of conflict from those of orderly 'succession of goals'. Indicative of such conflicts is the emergence of 'deviant' specialties, fields whose members 'accept the goals of their discipline, but believe their specialty is much more important relative to these goals than others give it credit for' ('reformers') or who 'in effect, reject the central goals of the larger discipline and, therefore, the legitimacy of the prestige system in it' ('rebels') (cf. Hagstrom, 1965, p.187).

In opposition to Hagstrom, the causes for 'reform' and 'rebellion', in Hagstrom's terminology, must be solved primarily from the level of cognitive orientations, and their interdependencies. Thus, it can be clarified with the help of the differentiation cited by Martins between 'restricted' and 'unrestricted' sciences (1972, p.78) that systematic connections within the structure of cognitive orientation complexes may be the reason for goal conflicts. Relativity theorists are a specialized and marginal group within physics compared to the quantum theorists; they are constituted as a group over a paradigm, the acceptance of which is limited with respect to the rest of the discipline, and it is evidently limited with respect to its potential for puzzles. This does not touch upon the fact that relativity theory has revolutionized modern physics and competes with quantum theory. The rejection of the stratification system in physics, especially with reference to their own field, that differentiates the relativity theorists from the solid state physicists, for example, may be traced back to the fact that their paradigm cannot be derived from that of quantum theory. Both groups are differently oriented on the level of 'sociological paradigms', the one empirically experimental, the other theoretically mathematical. Relativity theory is intended to do more than just solve a derived partial problem, namely the formulation of a unified field theory. Therefore the rejection of a stratification system which attributes only a minor role to it can hardly be surprising. That this does not have any consequences for the stratification of the entire discipline may be sought in the fact that relativity theory has not yet proved its superior explanatory power in terms of anomalies to be integrated and therefore is not institutionalized as a cohesive social organization (cf. Hagstrom, 1965, p.188ff).

Hagstrom's example of molecular biology points in the same direction, namely that the reasons for the rebellion have to be looked for on the cognitive level. The subdisciplinary boundaries of biology, between zoology, botany, physiology and bacteriology are blurred to the degree that in all these fields scientists have started to study life at the molecular level. The classical goal or, that is, the traditional conceptual systems (perhaps the metaparadigm ?) of biology are replaced by those of chemistry (cf. Hagstrom, 1965, p.193). Böhme, van den Daele and Krohn interpret the same process as an internalization into chemistry of the goal to study the biological cell with the result that a research process is induced which moves chemistry into the direction of molecular biology (Böhme, van den Daele, Krohn, 1972, p.39). The historical genesis of this development is decisive and the question what this change in goal means for which discipline, namely integration of a new goal on the one side or possibly replacement of a paradigm on the other. Independent of in which discipline the development of molecular biology originates, the reasons for its development must be sought in the structure of cognitive orientation complexes of biology and/or organic chemistry which is 'unrestricted' in so far as it makes it necessary to follow the analysis of problems beyond the disciplinary boundaries and into other disciplines. The result of the advance into the sub-cellular sphere has, as Hagstrom's citation of Barry Commoner documents, far-reaching consequences for biology, although not all traditional problems of its specialties have become obsolete. For the molecular biologist, however, the goal to analyse and explain the biological cell and the scientific achievements which are based on it can no longer be binding in the traditional sense. Molecular biology has become institutionalized and goal conflict between it and biology, which so far has only led to structural differentiation, may lead to a revolution as soon as it should become inevitable to recognize that 'life is a function . . . of a molecule and not of the cell' (cf. Hagstrom, 1965, p.194).

Against these developments induced on the cognitive level, resistance is building up on the institutional level which alone gives the terms 'reform' and 'rebellion' their meaning. Hagstrom's analysis of areas of strain such as university appointments, student training and opportunities of publication as well as their specific mechanisms of adaptation in view of goal conflicts is sufficiently instructive and does not have to be repeated (cf. Hagstrom, 1965, p.206ff).

Because the discoveries whose dissemination has been prevented on account of institutional conditions, despite their validity, are not recorded in the history of science, it is impossible to get a complete picture of the effects of institutional factors. Indicative cases, however, are the rediscoveries of originally rejected theories. Barber's analysis (1961) of the 'Resistance by scientists to scientific discovery', in which he makes 'cultural' ('substantive concepts', 'methodological conceptions', 'religious ideas') as well as 'social' ('professional standing', 'professional specialization', 'societies, schools, seniority') factors responsible for the resistance to innovations and discoveries, reveals the difficulty in distinguishing analytically conditions of the general 'cognitive situation' from the effects of their 'social substrate'.

The history of rediscoveries supports, however, the assumption that dominant orientations, paradigms and 'conceptual schemes' do not specify exactly what exists and how phenomena behave. The fact that cognitive orientation complexes do not sufficiently predetermine cognitive behaviour just as social norms do not entirely predetermine social behaviour is the reason that in science cognitive and social structure, although they analytically correspond to one another, do not need to in reality. Due to their limited independence, they may develop disparately. From rediscoveries, it may be indirectly deduced that 'independent' developments on the cognitive level do not succeed if cognitive orientation complexes and institutional structures of a discipline or a specialty are integrated, meaning that the cognitive consensus is high, finding its expression in a high degree of social institutionalization as well, so that cognitive and social structures correspond. Kuhn's anomalies may be taken to be the reverse case.

The same discoveries (or theories), that in a phase of 'normal science' are rejected and ignored, in crisis assume great importance, are accepted and replace the existing paradigms (cf. Kuhn, 1962, p.75). Kuhn only states the cognitive conditions for the recognition and acceptance of anomalies.

However, this account has to be expanded by an institutional dimension. Otherwise it is not entirely understandable why 'shocking discoveries' or newly 'invented' theories lead to crisis, why scientific élites are forced to surrender their norm explicating position of power or are overthrown in the course of revolution. Dominating orientation complexes are analytically the legitimate order of the institutional stratification system. Cognitive and institutional structures are integrated as long as the prevailing paradigms can solve, i.e. explain and integrate, the problems which appear in the cognitive process. However, as soon as 'counterinstances' appear, i.e. anomalies which can no longer be ignored, the 'problem solving capacity' of the paradigm, i.e. the capacity to account for the anomalies in its own frame of reference, and with it the legitimacy of the institutional order become doubtful. The resistance of scientific élites to the recognition of anomalies as well as to the discussion of fundamentals is a lost battle for the retainment of power which at this stage has already lost its epistemologically rational and legitimate basis. The revolution brings that group to power which is able to formulate a comprehensive paradigm that is superior to the old one, having a greater 'problem solving capacity', i.e. which can integrate the accumulated anomalies in a new context of interpretation.

How far-reaching this revolutionary change of power is depends on the specific relation of orientation complexes to anomalies. In contrast to processes of differentiation, however, it involves the replacement of old paradigms with new ones as well as the overthrow of the traditional élite and finally the replacement of the old stratification system. The unstable orientation to reputation collapses because reputation loses its legitimacy. In other words, its complexity-reducing function can only be reconstituted on the basis of a new paradigm. Where *élites* have lost their legitimacy and no longer determine the direction of research, the institutional mechanisms of power such as socialization, resource allocation and appointments must also collapse, be it with some delay. As long as the 'paramount values' or the political 'culture' of science retain their authority, prolonged phases of illegitimate power can hardly exist.

The examples of forms and sources of change in science given so far focus on the dynamics of the cognitive sphere. However, institutional structures of science are also changed intentionally and autonomously by governments with the goal of directing research into problem areas which would not be covered if science were left to the disciplinary logic of evolution, itself influenced considerably by the nineteenth-century university system in Germany (cf. Ben-David and Zloczower, 1962, p.49). Examples of such interventions could probably be found throughout the history of modern science. The novelty in this, though, can be seen in the fact that these interventions are being made on the basis of rational reflection upon the relation between 'social needs' and 'goals of science'. Although until now most instruments of science policy-making have been of a 'response-to-scientific-needs-and-opportunities' type and most of the political activities have been limited to the realm of 'applied science', this is changing.

'External' goals are becoming more complex, necessitating 'fundamental research' and thereby blurring the distinction between the two types of research altogether (cf. Bernal, 1967; Platt, 1969; Jantsch, 1971; Böhme *et al.*, 1972). 'Externally' initiated goals which are implemented by institutional measures such as the restructuring of universities, the foundation of particular research laboratories or a specific funding policy may or may not meet with resistance not only in the established institutional structure of science itself but also on the cognitive level. Not every social or political goal is rational (in the epistemological sense) or able to be pursued by particular fields of science on the basis of the respective body of knowledge. In other words, the cognitive orientation complexes serve as selection criteria determining the range of 'external' goals that may be integrated into science. They may change themselves, however, under the influence of 'autonomous' changes in the institutional structures. This is the process by which originally 'external' goals become internalized by the respective scientific field. In the *ex post* analysis they appear to be steps in the internal development of science, seemingly accounting for its internal logic of evolution.

Conclusion

The foregoing analysis is little more than a conceptual clarification and must be applied to concrete cases of scientific change to prove its fruitfulness. It seems justified to assume, however, that this approach, notwithstanding further clarification and improvement, itself avoids a number of common mistakes.

By taking the cognitive and the institutional level, their inherent connection and their own dynamics seriously, this approach does not have to resort to false dichotomies or to mutually exclusive mechanisms determining change such as the notions of a 'utilitarian competition' and 'epistemological rationality'. They may operate together and affect each other in a concrete situation without either one alone determining the process (cf. Kuhn, 1962, 1972).

Further, the presupposition of a certain 'concept of science', namely a strictly positivistic one, implying a cumulative growth is avoided. This concept of science, which seriously impairs many accounts of the evolution of science has to be abandoned. While analyses of scientific change based on this concept allow only for retardation or acceleration of scientific *growth*, and at most for processes of differentiation of scientific inquiry, it must be possible to include processes of revolution, redirection, or i.e. the substitution of goals in the evolution of science (cf. Kuhn, 1962, p.177). The approach presented here forces one to differentiate analytically between an *ex-post* reconstruction of the evolution of science and an analysis of the factors in an actual situation determining its course. The former leads to a teleological, the latter to a Darwinistic pattern of evolution (cf. Böhme *et al.*, 1972).

Finally, this approach which requires the study of processes of institutionalization of science, may produce insights of considerable practical relevance, namely illuminating the possibilities and limitations of a 'rational' steering of the development of science, 'rational' pertaining here to a policy which makes science serve social ends without either destroying science or becoming trapped in vain efforts.

References

Barber, Bernard (1961), 'Resistance by scientists to scientific discovery', *Science*, 134, pp.596–602.

Barnes, S. B. and Dolby, R. G. A. (1970), 'The scientific ethos: a deviant viewpoint', *European Journal of Sociology*, 11, pp.3–25.

Ben-David, J. and Zloczower, A. (1962), 'Universities and academic systems in modern societies', *European Journal of Sociology*, 3, pp.45–84.

Bernal, J. D. (1967), *The Social Function of Science*, M.I.T. Press.

Böhme, G., van den Daele, W. and Krohn, W. (1972), 'Alternativen in der Wissenschaft', *Zeitschrift für Soziologie*, 1, pp.302–16.

Cole, Stephen (1970), 'Professional standing and the reception of scientific discoveries', *American Journal of Sociology*, 76, pp.286–306.

Feyerabend, Paul (1970), 'Consolations for the specialist' in Lakatos, I. and Musgrave, Alan E., *Criticism and the Growth of Knowledge*.

Hagstrom, W. O. (1965), *The Scientific Community*, New York: Basic Books.

Jantsch, E. (1971), *Science and Human Purpose*, DFG: Forschungsplanung, Wiesbaden.

King, M. D. (1971), 'Reason, tradition and the progressiveness of science', *History and Theory*, 10, 1, pp.3–32.

Kuhn, T. S. (1962), *The Structure of Scientific Revolutions*, Chicago University Press.

Kuhn, T. S. (1963), 'The function of dogma in scientific research' in Crombie, A. C. (ed.), *Scientific Change*, London, pp.347–69.

Kuhn, T. S. (1970a), 'Postscript 1969' in Neurath, O., Carnap, R. and Morris, C. (eds), *Foundations of the Unity of Science*, II, Chicago University Press.

Kuhn, T. S. (1970b), 'Logic of discovery or psychology of research' in Lakatos, I. and Musgrave, Alan (eds), *Criticism and the Growth of Knowledge*, pp.1–23, Cambridge University Press.

Kuhn, T. S. (1970c), 'Reflections on my critics' in Lakatos, I. and Musgrave, Alan E., op. cit. pp.231–78.

Kuhn, T. S. (1972), 'Notes on Lakatos', *Boston Studies in the Philosophy of Science*, vol. 8.

Lakatos, I. and Musgrave, Alan E. (eds) (1970), *Criticism and the Growth of Knowledge*, Cambridge University Press.

Luhmann, N. (1968), 'Selbststeuerung der Wissenschaft', *Jahrbuch für Sozialwissenschaft*, 19.

Martins, H. (1972), 'The Kuhnian revolution and its implications for sociology' in Nossiter, T. *et al.* (eds), *Imagination and Precision in the Social Sciences*, London: Faber and Faber.

Masterman, M. (1970), 'The nature of a paradigm' in Lakatos, I. and Musgrave, Alan E. (eds), *Criticism and the Growth of Knowledge*, pp. 59–89.

Merton, Robert K. (1957), 'Science and democratic social structure' in Merton, Robert K., *Social Theory and Social Structure*, Chicago: Free Press.

Merton, Robert K. (1968), 'The Matthew effect in science', *Science*, 159, January.

Musgrave, Alan E. (1971), 'Kuhn's second thoughts', *British Journal of the Philosophy of Science*, 22, pp.287–97.

Parsons, Talcott and Shils, Edward A. (1962), 'Values, motives and systems of action' in Parsons and Shils (eds), *Toward a General Theory of Action*, New York: Harper & Row.

Platt, John (1969), 'What we must do', *Science*, 166, pp.1115–21.

Polanyi, M. (1958), *Personal Knowledge*, London: Routledge & Kegan Paul.

Polanyi, M. (1963), 'The potential theory of adsorption', *Science*, 141, pp. 1010–13.

Ravetz, J. R. (1971), *Scientific Knowledge and its Social Problems*, London: Oxford University Press.

Storer, N. W. (1966), *The Social System of Science*, New York: Holt, Rinehart and Winston.

Toulmin, S. (1970), 'Does the distinction between normal and revolutionary science hold water?' in Lakatos, I. and Musgrave, Alan E. (eds), pp. 39–47.

Whitley, Richard D. (1972), 'Black boxism and the sociology of science: a discussion of the major developments in the field' in Halmos, P. (ed.), *The Sociology of Science*, Sociological Review Monograph No. 18, Keele University.

Zuckerman, Harriet (1970), 'Stratification in American science', *Sociological Inquiry*, 40, pp.235–57.

4

Cognitive and social institutionalization of scientific specialties and research areas

Richard Whitley

Introduction

Recent work by Hagstrom (1970), Griffith and Mullins (1972) and others, has pointed to the significance of differences between scientific specialties for the formation of social groups and patterns of information transfer. However, no consistent nomenclature for the classification of cognitive structures and social groups has emerged and, in general, the cognitive development of scientific areas has been taken for granted as non-problematic (cf. Mullins, 1972). The social structure is assumed to correlate, at certain points, to the cognitive structure in the natural evolution of a scientific specialty without consideration of the possibility that the form of existing social and cognitive structures may affect new cognitive developments. By adopting the Kuhnian model of scientific development a linear and necessary evolution is assumed which begs the question of how different types of cognitive structure arise, develop and are maintained. In this way, the study of science is reduced to the study of how social structures match the ineluctable series of stages of cognitive development towards a monistic paradigm.

If this model is taken as problematic, then the question becomes: under what conditions does cognitive monism appear? To answer or even to be able to discuss different modes of cognitive development, some categories for analysing social and cognitive environments need to be explicated.

When more than one model of cognitive development in science exists (there are now at least two relatively well articulated ones, those of Kuhn and Masterman and of Popper, Lakatos and Feyerabend), a way of describing the different situations in which different models apply is required. That is to say, as soon as the development of scientific ideas is taken as problematic, some means of characterizing different sets of ideas and their structure and their social context is essential. To merely assert that biology is different from physics is not very informative unless we can say in what way they are different and how their differences can be expected to affect their development. This chapter will tentatively outline some categories for discussing differences in cognitive and social relations in scientific development.

In the main, I shall be concerned with a more micro level of analysis than is common in the area, with a lower level of cognitive structure than highly general 'paradigms'. It is largely because of my experience studying scientists in solid state physics where the categories used by philosophers and sociologists were of little use in explicating what the scientists were actually doing and how their work developed that I will concentrate on the micro level. It may be that the categories I outline are inapplicable outside physics, but in principle they are intended to refer to all scientific cognitive structures. In making sense of any cultural development, we must take into account the various implications of existing structures. To do this we must reconstruct their history, as Lakatos (1971) points out, but the sociological problem is to account for the particular manifestations which did occur, in the light of possible alternatives. We have to be able to discuss scientists' actions in the context of their cognitive and social environment. It is this process I am concerned with, my aim being to present possible categories for the analysis of such environments and their connections.

The major difference between scientific structures discussed in this paper is the degree of institutionalization. This factor is of considerable importance in relation to scientific development in that it expresses the degree of coherence and articulation of scientific ideas which, in turn, affects the extent to which scientists can develop within a particular framework and identifying points requiring alternatives. Rather than positing one fully articulated 'paradigm', as Kuhn does for 'mature' science, or a set of competing fully articulated theories as the Popperians tend to do, I am suggesting that science consists of a variety of cognitive structures of varying degrees of closure, coherence and articulation and that the mode of variation of these structures has consequences for their development. Furthermore, there are different levels of cognitive structure in science which may also vary. Differing degrees of institutionalization at different levels may be seen as constituting degrees of permeability to novelty, degree of resistance to alternative representations and interpretations of results or to alternative modes of understanding. By using the concept of institutionalization, differences in the rate of diffusion of types of ideas can be discussed for different cognitive enterprises.

The concept of institutionalization as I will use it refers primarily to the patterning of actions and meanings. The degree of coherence and organization of actions and perceptions and the extent to which ideas are articulated and adhered to constitute the degree of institutionalization. While this necessarily includes some commonality of expectations, it does not imply an integrated normative order which dominates actions. Common expectations at one level need not be integrated into a coherent system of value orientations at another. There may, for example, be a 'correct' way of using a piece of physical apparatus to obtain results without implying that everyone who uses that apparatus subscribes to the same model of the 'ideal' explanation or even interprets the results similarly. An area is highly institutionalized when the scientists share a common attitude concerning its aims, methods and explanation ideals. This attitude exists as a distinguishing characteristic of the area rather than as an unstructured amalgam of expectations from other areas. In general, there are two types of institutionalization in science: two aspects of scientific activity which are patterned to varying degrees, the cognitive and the social.

It should perhaps be emphasized that I do not intend a separation of the cognitive from the social where scientists' cognitive activity is seen as unconnected to their social actions. On the contrary, I am concerned with the different modes of connection between these aspects. Scientists are social beings and science is a social activity and scientific understanding is an expression of that activity. In distinguishing cognitive institutionalization from social institutionalization, I am suggesting it is fruitful to analyse differences in the extent of coherence and cohesion between intellectual products, their mode of production, and the social circumstances surrounding their production, evaluation and revision. By allowing for these differences, I do not mean that highly integrated social groups can be said to exist without a set of integrated cognitive meanings or that a systematic set of ideas held by a number of scientists does not imply some form of group association. I am suggesting that different levels of meaning can be integrated with different types of social structure and it is not necessarily the case that, for example, invisible college social formations are always associated with monistic, closed, complete 'paradigms' covering all aspects of scientific thought. An institutionalized cognitive structure may exist at the specialty level without necessarily implying a cohesive social structure at that level since research areas may be constituted by sectarian groups who concur on the basic tenets of, say, an observation theory, but who differ on its range of applicability.

Cognitive institutionalization has two major related aspects. First, it refers to the degree of consensus and clarity of formulation, criteria of problem relevance, definition and acceptability of solutions as well as the appropriate techniques used and instrumentation. Second, it defines the activity of a scientist in terms of the consensus. In an area of relatively high institutionalization, we can predict with a fair degree of accuracy what a scientist will be doing, which models he will use and what sort of 'ideal' explanations he will accept. Social institutionalization also has two dimensions: first, the degree of internal organization and boundary definition and second, the degree of integration into the prevailing social structures of legitimation and resource allocation. For science this second dimension usually refers to the degree of integration into university departments and teaching curricula.

A further distinction is useful in terms of the scope of the institutionalized activity. Again, a simple split is proposed between: (i) a set of similar problem situations and (ii) a general approach for the analysis of a number of such sets. The former, I term a research area and the latter a specialty (cf. Hagstrom, 1970, pp.91–2). These distinctions are meant to be orthogonal. There could be differing degrees of cognitive and social institutionalization in both research areas and specialties. The degree of similarity among problem situations may vary according to who is making the judgment and certain types of similarity may be more important than others in delineating the cognitive boundaries of a research area. What constitutes a problem situation will be discussed later, although I will not deal with how a situation comes to be seen as problematic. My concern is rather with characterizing alternative modes of development—cognitive and social—from an initial perception of a problem.

Cognitive institutionalization

In referring to a consensus regarding criteria for the definition and evaluation of scientific work, I intend something more than Mulkay's cognitive and technical norms (1970) which define as legitimate certain problems and techniques respectively. It is the articulation of such norms into a relatively coherent mode of understanding—a set of interrelated and interdependent, though not necessarily logically, statements about the world and how to study it—which is implied by this aspect of cognitive institutionalization. The greater the extent of such a process the more clearly defined are the criteria for evaluating scientific work and the more systematic and coherent the structure of understanding containing them. In the ideal this structure would presumably fulfil the expectations of certain philosophies of science, but here I am concerned only with the distinction between a situation where the criteria for evaluation are vague and loosely adhered to, and where little coherent, integrated understanding of the object of analysis is available and a situation where the criteria are clear, consistent, held in common and it is felt that a considerable degree of understanding has been achieved. Where there is an integrated mode of understanding, criteria for adding to that understanding will usually be evident. It should, perhaps, be emphasized that in the case of high institutionalization written, formal explication of the understanding achieved is not essential as long as a tradition of this understanding is maintained. Formal enshrinement in textbooks may follow, but the activities of practitioners of a field may be highly institutionalized much earlier. It is probably essential, though, that a commonality of tacit knowledge or, as Ravetz has it, 'craft' knowledge (1971), is achieved for full cognitive institutionalization. Without agreement on the low level technical knowledge for turning 'data' into 'information' and 'results', a coherent cognitive understanding is unlikely. It could be argued that a minimum of consensus on this low level of knowledge is a precondition for the existence of an empirical science.

Scientific ideas, or 'idea systems', then, are distinguished by their degree of articulation and commonality to a problem area. Cognitive institutionalization covers both intellectual coherence and order and the commitment to, and agreement on, this order. A low degree of cognitive institutionalization refers to a low degree of intellectual order and, correspondingly, little common commitment or agreement. In this situation, scientists will probably adhere to some common basic values and beliefs concerning the nature of the scientific enterprise and possibly some perceptual commonalities, but their work will be unconnected and disjointed. Common definitions and use of technical terms will probably be lacking and in so far as any linguistic or symbolic overlap occurs it will be through everyday language rather than through a relatively autonomous, specialized jargon. Intermediate between the two extremes is the situation of competing, clearly articulated idea systems explicating the same reality. Here, rival groups of scientists are fully committed to differing intellectual orders and sometimes these cognitive structures are incommensurable. More usually, while substantial divergences of interpretation and explanation may occur, total incommensurability is unlikely since idea systems are rarely fully and completely articulated. While it may be possible for the degree of commitment to vary independently of the degree of articulation of an idea system, it is unlikely that commitment to a set of ideas will occur without attempts at systematization. The converse does not hold. Many cognitive structures exist without current commitment being expressed. Furthermore, commitment to a set of techniques or observation theory may occur without corresponding commitment to explanatory models because there are no models perceived as adequate. In other words, the part of an idea system which enables research to continue may be highly articulated and influential, but other aspects much vaguer and unsatisfactory with competing interpretations. In such a case, high cognitive institutionalization would exist at one level, but not at another.

Another aspect of cognitive institutionalization is a corollary of the first. It is the exemplification of institutionalization in terms of an individual scientist's activity. It refers to the predictable nature of his work, given his definition of his field of interest. In an area of low institutionalization, his self-labelling will not be very informative about what he actually studies, or how, or what sort of understanding he wishes to reach, whereas in an institutionalized area a similar item of information may indicate which materials he is interested in, the instrumentation used and the representation of his results. In uninstitutionalized areas, the scientist will have to be much more specific about what he is doing, and why, to provide the same depth of information as occurs at a more general level in a more structured situation. Compare, for example, a solid state physicist with a cancer researcher. This variation in the predictability of a scientist's actions is related to the degree of cognitive identity, the extent to which he is able to identify his work as derived from a particular, articulated, mode of understanding. The more cognitively institutionalized is the area he is working in, the clearer he will be about his cognitive identity and it will be easier for him to distinguish 'his' area from others and the background knowledge behind his current problem situation from others. Institutionalization in this sense refers to the common ability to distinguish one cognitive structure from another and to allocate problems as belonging to one area rather than another. Such cognitive identification, of course, would be expected to lead to social identification which has implications for the internal social order of the area and the external recognition of its existence and allocation of resources for its continuance.

Social institutionalization

The existence of a coherent cognitive commonality in an area may be accompanied by a similarly structured social community, but not necessarily. Recognition of common cognitive interests and understanding may not always lead to social cohesion, but will probably lead to the development of social boundaries, although they may not become solidified and may disappear as cognitive developments continue.

Social institutionalization refers to the creation and maintenance of formal structures which demarcate members of a cognitive structure. In degree, it ranges from being aware of other scientists working in related areas and exchanging ideas or results to the full panoply of professional societies, core journals and codes of ethical conduct. In a similar manner to the growth of cognitive identity, socially institutionalized areas will provide the basis for social identity. Here, it will be relatively clear which professional societies a scientist will join, which meetings he will attend, which journals he will regularly examine and which organizational unit he will be located in and those he will have most contact with. He will be able to define his professional social circle and distinguish his social base from others. In contrast, socially uninstitutionalized areas will not have a clear structure of journals, professional meetings and organizational levels and social demarcation will be correspondingly difficult. In place of elaborate formal communication media and meetings which serve as demarcatory devices and bases for social identity, a loose array of overlapping arrangements will emerge with no obvious structure.

In this situation, scientists may form relatively small, fairly cohesive social groupings based around a common problem or mode of understanding as a means of coping with the lack of external structure. Personal contacts will be more important as a means of obtaining information and legitimation of one's work and others' research. While in highly socially institutionalized areas which have a high degree of cognitive consensus, journal referees will be able to apply consistent and coherent standards, in situations where there is low cognitive consensus evaluative criteria will differ and additional filtering mechanisms such as personal validation by trusted colleagues become necessary. Nevertheless, there is no necessary correlation between the degree of cognitive and social institutionalization as the synthesizing and systematization of a cognitive framework may take place without extensive social co-ordination. However, it is unlikely that once a 'central dogma' has emerged—even fortuitously—a social structure to ensure its continuity and application will not arise. It is possible that an elaborate social apparatus will exist in a field where cognitive developments have rendered the consensus meaningless and conflicting criteria are commonplace. Similarly, of course, a new cognitive structure may arise in a social framework which originally contained a totally different mode of understanding.

Social institutionalization in a societal sense may occur at a considerably later date than the initial structuring of the cognitive consensus. It is dependent on a number of factors which do not impinge so directly on the other processes. Formalization of an activity into the currently legitimated modes of social organization is only likely after a substantial degree of cognitive and internal social institutionalization has been achieved unless dominant sources of funds and legitimation insist on promoting a particular field. It is possible that by the time integration into formal organizations has occurred the label under which incorporation takes place has little connection with the actual cognitive structure being developed. Similarly, the social groups which constituted the basis for social institutionalization may have disappeared. Formal incorporation into a university department need not necessarily tell us anything about the actual work except that it bears some relation—with varying degrees of tenuousness—to a 'successful' cognitive enterprise. With regard to the current dominant mode of social organization—the university—we may note the following stages of institutionalization. First, some Ph.D. students are allowed to work in the area and obtain their certification. Second, a post in the department is reserved for a specialist. Third, a personal chair is created for a distinguished practitioner. Fourth, a subject chair is created. This movement can also be examined through the number of universities at each stage and the number of optional or required undergraduate examination papers in the area. In some departments, of course, most of the research centres on one particular area. This usually would not take place until the area was sufficiently respectable that the senior professor could follow such a specialization policy and be sufficiently influential to obtain research funds. The establishment of a separate panel to judge grant applications in the area is also a symptom of full institutionalization. This aspect of social institutionalization will not be considered further in this chapter since I am primarily concerned with more micro aspects of scientific development.

Research areas and specialties

The distinction between a scientific research area and specialty is partly one of scope and partly one of kind. A specialty may be considered as an agglomeration of research areas or a set of sets of problem situations. It also implies, however, a mode of understanding which structures and integrates different problem situations, e.g. field theory in plasma physics. Such integration is not necessarily by means of monistic models, or explanatory schema, but may occur through the common application of a particular instrument to a number of research areas. In this sense, there can be no totally uninstitutionalized specialty in that there must be some rule or criteria for relating sets of problem situations to each other if we are to speak of a specialty. The case of non-institutionalization is, therefore, logically impossible. It is possible, however, to speak of low and high degrees of institutionalization with regard to both research areas and specialties.

A research area has been characterized as a set of problem situations. The principles according to which these sets are ordered will vary. But, in terms of subjective perceptions of cognitive commonalities, we can briefly outline the main possibilities. First, the phenomena under investigation may be similar. The definitions need not be identical for considerable commonality. Superconductivity, for example, may be studied in a number of materials using different techniques, but there is agreement on the essential characteristics of the phenomenon and different sets of observation statements and postulated mechanisms have extensive overlap. Second, a material or 'system' may form the basis of common problem situations. An illustration is the study of amorphous materials and their theoretically surprising behaviour under certain conditions. These non-crystalline solids have been found to exhibit certain magnetic and electrical phenomena which are anomalous in terms of the prevailing model in solid state physics. A number of different techniques are being used to study the various anomalous phenomena and the set is defined by the characteristic of the materials being non-crystalline. At a lower level, this form of commonality may occur in relation to a single system or material, such as one of the rare earth metals, although it is usually taken to be representative of a particular set of systems which exhibit 'interesting' features. Interesting, of course, in terms of the dominant model which means that research areas based on peculiarities of materials tend to occur only when well articulated explanatory and descriptive mechanisms have been developed. Third, the common use of particular instrumentation with its associated rules for obtaining meaningful information acts as an ordering principle. This usually occurs when the technique is comparatively complex and its use requires expertise. The acquisition of the cognitive and technical skills necessary to operate a complex piece of apparatus may require lengthy training; thus, it acts as a demarcation criterion for the research area. Techniques based on reactor physics are a somewhat extreme case (cf. Frost and Whitley, 1972), but other examples of boundary criterion are the use of liquid helium in low temperature physics or radiation techniques in radiobiology. The use of techniques as a demarcation criteria will vary according to their novelty as well as their complexity. Initially, the introduction of neutron beams and electron microscopes to the analysis of biological tissues may form new research areas, but should a consensus be reached on how to interpret data obtained by these instruments and their use become commonplace, then research areas based on phenomena or materials may emerge—either within the general context of the technique which will then become the organizing principle of a specialty—or using a number of techniques. Of course, in practice, various combinations of these organizing criteria may occur. Pure types are unlikely to be numerous.

The objects of these criteria or principles—problem situations—are the immediate problematic of the scientist. What he is concerned to understand, the area of uncertainty within the cognitive boundaries he delineates, constitutes his problem situation (cf. Ravetz, 1971, pp.132–40). In any concrete instance, a scientist may well have more than one problem simultaneously which may or may not be directly related, but he is usually able to state the immediate area of concern and sometimes identify the larger problem. The way in which he sets the boundaries to his problem situation and the recognition he extends to others working on similar problems indicate the organizing principle(s) of the research area. It is possible, of course, that no recognition is extended and each scientist considers himself isolated even though, to others, there exists some basis for cognitive commonality across problem situations. While we may speak of a research area in such situations in that we, or some other external observer, can use an organizing principle to link different scientists' problem situations, it would be premature to assert the scientists' irrationality in failing to recognize common features without full investigation of the circumstances. One reason for non-recognition in practice may be the considerable degree of impermeability obtaining in discrete cognitive structures which are relevant to the particular problem. In other words, while it may appear that a number of scientists are working on similar problem situations, they may be using separate techniques or explanatory models which are incompatible and so do not see each other as research colleagues.

Specialties are more general in scope than research areas and organized, to varying degrees, around particular ways of understanding the world not so much at the 'negative heuristic' level (cf. Lakatos, 1970, pp.133–4) as at the 'positive heuristic' and observation theory levels (Lakatos, 1970, pp.134–8). As mentioned, a particularly complex technique may constitute the basis for a specialty because it provides a system for investigating a set of problems and evaluating solutions. High voltage electron microscopy might constitute such a basis. In a sense, of course, the simplest technique implies a complex set of theories in producing information; for example, an optical microscope entails a theory of light which may be highly problematic (cf. Feyerabend, 1970), but this is often relegated to the category of 'background knowledge' which is assumed as non-problematic for the immediate purposes and will only become a source of concern if substantial difficulties are encountered. When a technique becomes so accepted that it is generally included in the scientists' basic intellectual equipment without further thought, it ceases to be a meaningful demarcatory device for current work, although as the basis for organizational titles it may flourish. In general, specialties are built around a set of cognitive structures which order and interpret a particular, restricted aspect of reality. Such cognitive structures are often termed models, sometimes analogies or metaphors (Achinstein, 1968; Hesse, 1966, 1968). I do not here intend to discuss the growing literature on the role and types of models in science, and will use the term broadly as Harré (1970, pp.33–62) suggests; that is, as a mechanism which seeks to provide an understanding of how observable events are related, the inside of a 'translucid box' (Bunge, 1963, 1964; Whitley, 1972). This organizing principle is different in kind from that which forms research areas. I would argue that while problem situations are related by common perceptions of cognitive uncertainty of objects, compounds or techniques for study to constitute research areas, e.g. electron spin densities in rare earth metal compounds examined with neutron beams, the integration of the latter occurs by means of adoption of a common mode of understanding. In other words, some consensus exists on a model or models of the reality under investigation, for example, crystalline atomic structures. When a complex technique forms the basis of such an integration, it does so by implying, although perhaps not explicitly, some model by which data become meaningful results. Specialties, then, are distinguished by containing one model, or a limited set of models, which purport to explicate existing 'facts' and direct further investigation.

Cognitive and social institutionalization of research areas

Cognitive institutionalization of a research area refers to the firmness with which problem situations are connected and boundaries drawn between sets.

Whatever the organizing principle—or particular combination of different principles—the degree of agreement regarding its meaning, appropriateness and application is a major element in the institutionalization of the research area. A research area can be said to exist when scientists concur on the nature of the uncertainty common to a set of problem situations. Some agreement on what is problematic and what can be assumed is essential. The more consensus on the definition of phenomena, application of appropriate techniques and meaning of results, the higher the degree of cognitive institutionalization. In a situation of low cognitive institutionalization, there is some agreement on the common uncertainty in a set of problem situations, but there are conflicting views on how to deal with this and how to evaluate possible solutions. While there is some common use of a technique or common definition of a phenomenon, such as electrical leakage in a nerve, differences exist over research strategies and tactics. Depending on how a scientist may choose to define his research, i.e. the uncertainty he is concerned with, he may 'belong' to different areas.

A further aspect of cognitive institutionalization is the degree to which the organizing principle actually structures the activity of scientists so that there is a restricted, finite number of activities a member of a particular research area could be doing. Obviously, a specific machine defines some research area which precludes employing other instruments and may, more importantly, determine how measurements are taken and interpreted. A situation may also occur where common materials are the defining property of a research area with a specific set of operations to be carried out to examine the phenomena because of their particular characteristics—considerable instability at certain temperatures, for example. Similarly, the more closely the phenomenon is defined as an object of investigation it may be that only a limited number of techniques and materials are suitable and then only in particular ways. For example, superconductivity only occurs under certain temperature conditions, and only for certain metals, and the so-called 'mixed' state between superconducting and normal occurs in particular alloys under particular magnetic conditions. The most constricted research area would be one where the area of uncertainty was so tightly delineated that only one material was being studied for one phenomenon and by only one technique. In that case, cognitive development would most likely involve minor modifications in the applications of the techniques together, perhaps, with movement to the examination of related elements. Such a research area would be highly cognitively institutionalized and the possible cognitive developments severely limited in the sense that its boundaries of cognitive uncertainty are unambiguously defined and modes of resolving the uncertainty clearly specified. These developments might, however, produce data which required radical restructuring of the problem situation and the organizing principle(s) of the research area. If this process continued, it would throw doubt on the dominant model and might lead to a restructuring of the specialty. A more likely response would be to divide the research area and reinterpret the model to deal with the recalcitrant phenomenon (cf. Hagstrom, 1965). In a highly institutionalized research area which was part of a highly institutionalized specialty, i.e. where the dominant model in the specialty demarcated research areas, it is unlikely the dominant model—or set of connected models—would be seriously reconsidered unless it broke down in a number of research areas after a variety of techniques and materials had been tried. The model of crystalline solid states, for example, is not currently under attack because of particular magnetic properties of amorphous materials, but is being extended and modified to account for these 'anomalies'.

The internal social institutionalization of a research area is largely predicated upon the common recognition of other scientists' work being directly related to one's own. This recognition may lead to one of three types of social organization. First, co-operation and division of labour within a cognitive consensus. Accordingly, scientists agree on a common approach to a particular set of problem situations and divide up the work so that each group has a sub-set of problems—or techniques—and deals only with those. Second, they may compete within a cognitive consensus. A general approach may be accepted with some consensus on the 'relevant' and 'interesting' problems and how to evaluate rival solutions, but different social groups work on the same problem situations with similar techniques and compete. Both of these types of social organization imply a fair degree of cognitive institutionalization of the research area which, in turn, implies the existence of articulated models, although the specialty need not be highly cognitively institutionalized since there may be competing models. The third category occurs when scientists compete in a situation of low cognitive consensus. By definition, there will be some agreement on the basic problem under consideration, but there will be marked differences about how to tackle it and what solutions are acceptable. Each group is characterized by a different technique or an explanatory model or both. There may be some overlap in observation theories with only certain evidence generally agreed as admissible. The intensity of the debate will depend on the clarity and coherence of the competing approaches and on the existence of some overall 'idea of nature' (Collingwood, 1965, pp.1–3). Consensus at the metamodel level, that of the discipline, enables agreement to be reached eventually, if only in terms of very general values. Once such social pluralism has become established, it will be correspondingly difficult to develop cognitive monism or even commensurability between interpretations of rival models.

The degree of social institutionalization is the extent of informal mechanisms for control between members of the research area and the formation of solidaristic groups within the area for particular activities. If scientists working on a common set of problems do not recognize each other as colleagues, there is zero degree of social institutionalization. The alternate extreme would occur when the research area is organized like a religious sect (cf. Robertson, 1970, pp.116–42) with comprehensive indoctrination, social exclusivity, limited recruitment and highly structured patterns of social interaction. If a research area was fully socially institutionalized, we would expect a parallel degree of cognitive institutionalization. In general, nevertheless, there is no necessary corollary between the degree of the two types of institutionalization. Differing degrees of social institutionalization ranging from routine exchange of reprints to highly organized continuous contact within a small number of institutions may be associated with varying degrees of cognitive organization, although there will be certain threshold points at which one aspect cannot occur without proximate location of the other. It is unlikely, for example, that a low degree of cognitive institutionalization would occur with a high degree of social institutionalization throughout a research area.

The reverse case is possible if the specialty is highly socially organized. A minimum degree of cognitive consensus is necessary for interaction to occur on a patterned basis among members of a research area, although the degree of social organization necessary to ensure systematic development of a set of problem situations is unclear. The impact of large-scale technology, necessitating division of labour and collaboration, on patterns of social interaction in science has been noted (Frost and Whitley, 1972; Gaston, 1969), but it is doubtful if cohesive 'invisible college' type social formations are essential to the growth of all new research areas (cf. Griffith and Mullins, 1972). I would suggest that this depends largely upon the social and cognitive condition of the surrounding specialties. If they are highly organized and in opposition to the embryonic area, it will have to 'close ranks' to survive, otherwise not. If they are pluralistic and 'open', new research areas will arise without much difficulty and a strong sense of cognitive and social identity is not needed to maintain them. Overlapping social circles will probably be the major characteristic of research areas in pluralistic specialties unless cognitive boundaries are impermeable and opposed.

The possibility of competition among scientists in a research area characterized by a low degree of cognitive consensus raises the question of the level of analysis research areas are imputed. If they are primarily a cognitive category, how can they be further cognitively subdivided in a state of cognitive 'pluralism'? The point is that although definitions of problem situations, and so criteria for relating them, are linked to particular interpretations of models and research approaches, scientists are able to agree about what is related while differing on how to approach each problem situation. Naturally, it would be expected that the more scientists concurred on definitions of research areas and on a number of dimensions, the more likely it would be that they would concur on the relevant approaches to the study of problem situations. But, provided there is some minimal agreement on categories relating problem situations, so a research area can be said to exist, consensus concerning methods of approach may vary.

Agreement on what is problematic across different problem situations is necessary for a research area to be distinguished, but a shared core of uncertainty need not imply common modes of resolution. The existence of overlap in definitions of problem situations does not necessitate overlap in evaluative criteria for solutions. Social competition between groups in a situation of low cognitive institutionalization results in a low degree of social institutionalization in the research area as a whole, but a high degree of social institutionalization within the sub-groups. Such a research area is socially pluralistic and internally cognitively differentiated with respect to research strategies and evaluations. Social competition in a situation of high cognitive institutionalization, on the other hand, implies a fairly high degree of social integration at the research area level and rivalry is limited in intensity and scope although this social integration is unlikely to become sectarian unless, as indicated, the cognitive consensus is so divergent from specialty and discipline values that it becomes consciously deviant and intensely opposed. This implies, of course, that there is a considerable degree of cognitive institutionalization at the specialty level. A slightly lower degree of social integration, but still considerable, may occur when the specialty is low in cognitive institutionalization and there are no clear modes of understanding available and research strategies are vague. The high degree of cognitive uncertainty may result in diffuse and cohesive social groups forming, particularly if the specialty is new and lacks legitimation from the discipline and the society.

Cognitive and social institutionalization of specialties

As mentioned earlier, specialties are delineated by containing models which serve to make sense of a particular aspect of reality and indicate cognitive developments. The degree of cognitive institutionalization in a specialty may be interpreted as the degree of both consensus over appropriate models and their application and the coherence of the models. Highly institutionalized specialties are monistic in the sense that one particular systematized model emerges as the dominant approach of the specialty, e.g. the crystalline model in solid state physics. A specialty which is only institutionalized to a limited extent may have a number of loosely defined models or one subject to different interpretations. It is possible to have clear models and weakly defined specialties when there is no consensus over where which model should be used and boundaries between models are not clearly defined in their range of application or relevance.

While research areas are sets of problem situations with a common core of uncertainty delineated by the application of models, specialties are cognitive units dealing with a particular aspect of reality. Just as research areas do not exist where competing models are so incommensurate that no agreement on what is to be assumed and what is problematic is possible between models, i.e. problem situations have no overlap of definitions of uncertainty, specialties do not exist where competing models so totally and exclusively define the reality they purport to explicate that these definitions do not overlap. Specialties, then, may be distinguished in terms of the object of competing modes of understanding. In so far as scientists use models purporting to explicate similar realities they are members of the same specialism. The extent to which they agree on the appropriate mode of understanding and the extent of its articulation and elaboration constitute the degree of cognitive institutionalization. Cognitive pluralism of specialties refers to competing models of a reality. Cognitive pluralism of research areas refers to competing applications of models, or interpretations of the same model, to a core of uncertainty within that reality.

The degree of consensus on the definition of the reality and the appropriate mode of understanding is related to the scientist's identification of 'his' specialty. Scientists who are uncertain how their problem situations are related to conflicting definitions of various realities and which models are appropriate where, are members of uninstitutionalized specialties. This situation can be described as highly unstructured. The field or discipline is defined by the set of very general assumptions shared by its members—Lakatos's 'negative heuristic' or 'irrefutable core' at the heart of the scientific enterprise, e.g. the notions of information and control in biology, but there is no generally agreed system for ordering experience below this level except in the research areas where presumably some consensus is achieved on problem definition and technique application in that research continues with scientists using a variety of approaches, articulated with different degrees of clarity, at different times and places with possibly an acrimonious debate over their relative merits. The specialties are characterized, here, by competing modes of understanding about a particular aspect of reality which is not clearly delineated. As long as there is some agreement about the object of general concern, e.g. the structure of solids or the behaviour of viruses or genes, and one or more explanatory mechanisms, a specialty is said to exist and cognitive institutionalization deals with the degree of articulation of possible models for exploiting this object, the number of such models and agreement on a ranking of them.

In so far as one model is deemed best for understanding, the general object and research areas are clearly differentiated in terms of a common interpretation of that model, i.e. areas of uncertainty are defined by the one approach, both specialty and research areas are highly cognitively institutionalized. However, it is not necessarily the case that agreement on a model at the specialty level implies agreement on how to apply that model to define problem situations at the research area level, although members of the same institutionalized specialty will be more likely to agree at the research area level than when the specialty is relatively uninstitutionalized. It should, perhaps, be emphasized that this process of institutionalization is not regarded as inevitable in all areas of science and, especially, I see no reason to suggest that monistic, highly institutionalized specialties are any more 'mature' or 'scientific' than others. There is no inevitable tendency towards higher institutionalization in all specialties and it would be unduly restrictive to insist on a linear model of cognitive development leading to greater consensus and 'normality' in all situations. The cognitive and social conditions under which specialty cognitive institutionalization takes place are problematic and await investigation.

Social institutionalization of specialties refers to the formal organization of communication and membership and the demarcation of one specialty from another. Again, a minimum of cognitive consensus is necessary for social institutionalization to take place although the full array of journals and professional meetings is scarcely essential for the formation of a consensus on a coherent set of ideas. As as well having distinct journals and sections in professional societies, specialties differ in the extent to which journals and meetings are seen as more important, more central to the specialty than others. A small number of 'core' journals may dominate the specialty or there may be little discrimination between journals in terms of importance. Similarly, where cognitive institutionalization is low, differentiation between journals in terms of subject matter or approach may be difficult and extensive literature searches, and discussion with colleagues, necessary to obtain needed information. In general, I suggest that where a specialty's journal information media are poorly differentiated, personal contact will become increasingly important as a filtering and legitimatory device. Some sort of 'invisible college' type of social structure may emerge in research areas in conditions of cognitive uncertainty at the specialty level without necessarily being associated with a distinct cognitive structure consciously developed in opposition to the dominant model (cf. Griffith and Mullins, 1972). This will be assisted if work is concentrated in a small number of organizations linked by a common technique as in audiology (Griffith and Miller, 1970). Otherwise, while the main unit of cognitive and social identity is more likely to be the research area than the specialty, cohesive social groups which are mutually exclusive are not likely to occur. Divergences of techniques, evaluation criteria and models mean that scientists may 'belong' to a number of research areas and so although they may be able to identify their major area, its permeability enables them to have a wide range of personal contacts covering a number of areas (cf. Mullins, 1966). Ability to differentiate problems and current modes of solution, thus research areas, may be combined with a wide social circle with the high cognitive uncertainty at the specialty level necessitating personal contact and the proliferation of models and techniques enabling these to take place across research area boundaries. In this situation, a low degree of social and cognitive institutionalization at the specialty level is associated with a moderate degree of cognitive institutionalization and a low degree of social institutionalization at the research area level.

Interaction between specialties and research areas

Two types, and two objects, of institutionalization have been delineated. Next, I will outline some patterns of interaction between these categories, particularly the relation between research areas and specialties. Earlier, I suggested that they could be regarded as orthogonal. However, it seems likely that certain combinations of types of research area and specialty will be most common. A highly cognitively and socially institutionalized specialty will probably not occur with many loosely defined and vaguely associated problem situations, rather its research areas will be fairly clearly delineated, criteria for problem definition and evaluation of solutions will be agreed upon and difficulties seen as interesting 'irritants' to the dominant model which may necessitate minor alteration in interpretation, but by no means constitute a substantive challenge. Where the specialty is highly cognitively and socially integrated, it is unlikely that research areas will be highly socially institutionalized in the sense of forming closely knit, cohesive groups; overlapping, specific, social circles are more likely. This does not mean that all research areas related to the dominant model will be cognitively cohesive, but that there is more likely to be cognitive consensus within an institutionalized specialty at the problem situation level than not.

In this situation, we would expect 'core' journals to be clearly differentiated, considerable formal organization of professional meetings and little evidence of solidaristic groups with their own mutually exclusive cognitive structures. Additionally, such specialties will be highly institutionalized into the dominant modes of societal legitimation and organization. Cognitive development will take place chiefly through extension of the model to new research areas. This process may eventually result in the dissolution of the specialty as a cognitive structure as the model becomes over-extended as Masterman suggests (1970). The continued application of the model to new problem situations will result in 'anomalies', problems which should be soluble but are not. Redefining these problem sets and forming new research areas by reinterpreting the dominant model may temporarily resolve the difficulties, but as the model is pushed further it breaks down more often and the continued creation of 'anomalies' at the periphery eventually calls into question the utility of the model throughout the specialty. Success, thus, breeds failure, although the higher the degree of social institutionalization, the longer this process may take for a given set of models. Alternatively, problem situations which are not clearly relevant to any institutionalized specialty may develop by interpreting models from different specialties and, if 'successful', attempt to extend this interpretation to other areas. This process will be affected by the degree of cognitive and social institutionalization of the specialties and research areas and their permeability to alternative ideas. Where a set of research areas define their problem situations in terms of a common approach, and use a common model which is derived from that in an institutionalized specialty for tackling these situations, it is unlikely that alternative definitions will be acceptable unless widespread breakdown is incipient.

In a situation where a specialty is not highly cognitively or socially institutionalized, research areas may become more important as sources of cognitive and social identity for scientists. Considerable cognitive uncertainty at the specialty level may be associated with a fair degree of cognitive or social institutionalization, or both, at the research area level. This may be substantial when throughout the discipline there is little specialty structure, either cognitive or social, and a number of new technologies are being introduced which provide new data and conflicting interpretations. Here, scientists may define their work in terms of specific sets of problem situations, rather than by reference to a particular well-articulated model applying to a number of problem situations. They will also be highly integrated into a social group based on a set of common definitions of problem situations. The fluidity of evaluation criteria and apparent equality of alternative competing models may also result in different groups of scientists in research areas adhering to distinct models and regarding other approaches as 'unscientific' or 'irrelevant'. While these competing groups might agree on what the basic problem is, their interpretations will differ as will their evaluation of results. Where social groups become extended in separate universities and 'schools' are created, any cognitive consensus is unlikely to form except perhaps through a radical redefinition of the specialty and research area (cf. Krantz, 1969).

This case is discouraged in an area possessing a common technology and observation theory. Totally opposed scientific groups, each with their own millenaristic ideology, will not characterize research areas where technological facilities are limited and collaboration is required to obtain access. If different groups are able to develop their own particular technique for analysis and instrumentation, it is correspondingly more likely that they will remain antithetical and unlikely that discourse across cognitive and social boundaries will occur. It should perhaps be noted that a plurality of techniques does not necessarily imply such a situation. Where some specialty institutionalization exists, scientists will use a number of different techniques as a way of corroborating results which appear anomalous and not as a means of demarcating mutually exclusive cognitive structures and social groups. 'Revolutionary' groups will, I suggest, only characterize a research area, or set of research areas, when there is no specialty structure to order the proliferation of competing modes of interpreting experience and each mode is allied with a particular analytic technique and instrumentation. The conditions under which such breakdown of specialty structures occurs, or alternatively, if a state of chaos is assumed natural, the conditions under which some cognitive order is maintained, remain indeterminate. Kuhn (1970) and Masterman (1970), among others, have suggested we consider paradigm breakdown occurring as a result of overstretching analogies, but if scientific development is seen as consisting of developing analogies, then presumably it leads to other results as well as to cognitive anarchy. As Masterman (1970) hints, I do not regard 'crisis' science as likely unless different techniques and instruments become allied to competing modes of interpretation and so observation theories are incommensurate.

Where a specialty is relatively uninstitutionalized there is little general agreement on which models, and how they are defined, to use and develop in studying a general problem, research areas will be defined according to various criteria and will often use competing models to deal with their problem situations. No overall approach will be directly inferable from the results produced, although since the models are unlikely to be totally incommensurate some understanding may be possible at the general level. Probably different definitions of problem situations will be derived from different interpretative models and research areas will be differentiated by those models. The degree of cognitive closure of the models will determine the intensity of competition and awareness of a common concern. A plurality of models and evaluative criteria implies a plurality of problem definitions and solutions which will lead to certain *ad hoc* means of differentiating criteria and these will become institutionalized as distinct research areas. Although these research areas will not necessarily be totally compartmentalized and some commonality will occur, their boundaries will be less permeable with respect to each other in this situation than in specialties which are highly institutionalized. While in uninstitutionalized specialties each cognitively institutionalized research area is characterized by its own language, when a dominant model exists a common language exists and research areas are demarcated in terms of that language, not by different languages. However, it should be remembered that the common use of a particular instrument provides the basis for a common language across cognitive barriers and in this case research areas are unlikely to become mutually exclusive.

Specialties, research areas and cognitive development

In highly institutionalized specialties with clearly delineated research areas, I have suggested that cognitive development occurs generally in the direction of extending the dominant model to new materials or by using new modes of analysis to develop greater understanding of previously defined phenomena. If we accept Bunge's (1963, 1964) distinction between 'black-box' and 'translucid-box' science, we may say that probing further within the translucid box represents this form of development while constructing new hypothetical mechanisms or being forced to radically alter the supposed underlying structure by inadequacy of existing models represents cognitive development of a less structured order. The latter form of cognitive development is likely to be less frequent, I suggest, in highly cognitively institutionalized specialties, particularly if they are also highly socially institutionalized, although it may occur in research areas at the periphery of the specialty in that the model is not designed to deal with these problem situations, but is somewhat 'stretched' (cf. Masterman, 1970). In general, it seems that the modification of Kuhn's approach suggested by Masterman is most applicable to radical innovation in highly institutionalized specialties, though the term 'revolution' must be used with great care (cf. Martins, 1972).

However, it is doubtful if such highly socially and cognitively institutionalized specialties are representative of all activities we term 'scientific'. By whatever criteria that honorific title is bestowed, it seems possible that the statistical norm of specialty structure may prove rather less institutionalized than, say, solid state physics or inorganic chemistry. The difficulty that biologists, in particular, have in naming their specialty attests to the ambiguity of cognitive and social structures in this field (cf. Mullins, 1966; Hagstrom, 1970) and may well occur in others. Here, a multiplicity of models, techniques and contradictory results is to be expected and controversy about relatively fundamental points endemic. Cognitive development may take the form of an agreed 'negative heuristic', or basic view of science and the world, with frequent formulation and extension of alternative 'positive heuristics' which direct research along particular paths. While new models may occur more often where there is not a very high degree of social institutionalization, if it is very low there may be little incentive to share new ideas. In any case, colleagues may not be recognized and possible alternative solutions not collectively discussed. The amount of background knowledge in these specialties which can be assumed will be less than in the highly institutionalized specialties, and uncertainty about the appropriate evaluative criteria correspondingly higher. This uncertainty may lead to an ideology of 'black boxism' with the system under investigation regarded as too difficult to examine and correlations between inputs and outputs accepted as constituting knowledge. This may occur especially when external social pressure demands 'results'.

In specialties with a relatively low degree of cognitive and social institutionalization, then, scientific energies may well be equally distributed between applying and deepening one particular 'translucid box' model and developing alternative definitions of, and approaches to, the problem. Alternative modes of transforming 'data' into 'information' and 'results' will occur within a problem area and the specialty may be defined simply in terms of a very general world view or a variously defined problem such as cancer or the origin of the universe. While at the specialty level, cognitive development will consist largely of reformulating and articulating different models and their implications for understanding some aspect of reality, at the research area level in this situation two alternative modes of development may take place. If the research area is highly cognitively and socially institutionalized, considerable consensus and clarity exist concerning the area of uncertainty and the assumptions which can be taken for granted. Here, development will consist of examining more materials with existing techniques or applying a range of techniques to a recalcitrant material. It will be based on the craft knowledge which forms the basis of the common expertise. Although it may not be possible to make much sense, at the specialty level, of what is found, the research area is sufficiently clearly defined for systematic, integrated research to continue. In a situation where this is not the case, either cognitive pluralism is combined with social sectarian groupings or cognitive boundaries are weakly defined and social groups not based on clear cognitive orderings, cognitive development takes one of two forms. These are either increasingly smaller, closed systems of incommensurable results and interpretations or disconnected wanderings across the cognitive map by atomistic individuals. In other words, in specialties and research areas which have a low degree of cognitive and social institutionalization, a myriad of ideas and data are produced by very small collectivities which are not connected and not necessarily consistent over time. The area will probably be continuously redefined as interests change and any striving for coherence will probably take place at a philosophical level rather than in the form of particular theoretical models.

Conclusions

In this chapter, I have elaborated some distinctions for differentiating social and cognitive aspects of scientific work for an analysis of the interaction of social and cognitive factors in scientific development. I have suggested that the degree of institutionalization is a useful concept for talking about scientists' activities and that cognizance must be taken of the different levels at which these activities occur. In particular, cognitive development at one level of analysis, the research area, is not necessarily the same as that at another level, the specialty, but these levels interpenetrate. Two assumptions are basic to this approach. First, science is essentially pluralistic, there are many cognitive structures which are not identical and do not develop identically. Second, scientists' cognitive structures operate on different planes. This does not imply that they are hierarchical in that changes in the 'higher' structures necessarily determine changes in the 'lower'. While connections between these planes do exist they are not usually of an implicative nature and are often tacit rather than formally expressed.

The next problem is to see how fruitful these categories are. A major difficulty occurs with boundary definitions and inevitably some arbitrariness is necessary. Furthermore, scientists' identification of their research areas and specialties—which necessitates their understanding the categories—must be integrated in the light of an analysis of the literature. Since the degree of institutionalization is taken to be problematic, the categories have to be applied by the observer on the basis of his understanding of the cognitive structures. This requires much more detailed elaboration of the terms and their mode of application than was possible here as well, of course, as knowledge of the scientific content and its history. The study of case histories using these categories could form the basis of a comparative understanding of scientific developments (cf. Small, 1972) and it is with this in mind that they have been developed. Whether they do prove to be useful depends on future research, but the need for some such set of terms seems undeniable.

References

Achinstein, P. (1968), *Concepts of Science*, Baltimore: John Hopkins University Press.

Bunge, M. (1963), 'A general black box theory', *Philosophy of Science*, 30, pp.346–58.

Bunge, M. (1964), 'Phenomenological theories' in Bunge, M. (ed.), *The Critical Approach to Science and Philosophy*, London: Collier-Macmillan.

Collingwood, R. G. (1965), *The Idea of Nature*, London: Oxford University Press.

Feyerabend, P. K. (1970), 'Problems of empiricism, II' in Colodny, R. G. (ed.), *The Nature and Functions of Scientific Theories*, Pittsburgh University Press.

Frost, Penelope A. and Whitley, Richard D. (1972), 'Technology, cognitive structure and the formation of social groups in scientific research', unpublished paper, Manchester University.

Gaston, J. C. (1969), *Big Science in Britain: A Sociological Study of the High Energy Physics Community*, Ph.D. thesis, Yale University.

Griffith, B. C. and Miller, A. J. (1970), 'Networks of informal communication among scientifically productive scientists' in Nelson, C. E. and Pollock, D. K. (eds), *Communication Among Scientists and Engineers*, Lexington, Mass.: D. C. Heath.

Griffith, B. C. and Mullins, N. C. (1972), 'Highly coherent social groupings in scientific change', unpublished paper, Drexel University, Philadelphia.

Hagstrom, W. O. (1965), *The Scientific Community*, New York: Basic Books.

Hagstrom, W. O. (1970), 'Factors related to the use of different modes of publishing research in four scientific fields', in Nelson, C.E. and Pollock, D. K. (eds), *Communication Among Scientists and Engineers*, Lexington, Mass.: D. C. Heath.

Harré, R. (1970), *The Principles of Scientific Thinking*, London: Macmillan.

Hesse, M. (1966), *Models and Analogies in Science*, University of Notre-Dame Press.

Hesse, M. (1968), 'The explanatory function of metaphor' in Bar-Hillel, Y. (ed.), *Logic, Methodology and Philosophy of Science*, Amsterdam: North Holland.

Krantz, D. L. (ed.) (1969), *Schools of Psychology*, New York: Appleton-Century-Crofts.

Kuhn, T. S. (1970), 'Logic of discovery or psychology of research' and 'Reflections on my critics' in Lakatos, I. and Musgrave, Alan E. (eds), *Criticism and the Growth of Knowledge*, Cambridge University Press.

Lakatos, I. (1970), 'Falsification and the methodology of scientific research programmes' in Lakatos, I. and Musgrave, Alan E. (eds), *Criticism and the Growth of Knowledge*, Cambridge University Press.

Lakatos, I. (1971), 'History of science and its rational reconstructions' in Cohen, R. S. and Buck, R. (eds), *Boston Studies in the Philosophy of Science*, VIII, Dordrecht: Reidel.

Martins, H. (1972), 'The Kuhnian "Revolution" and its implication for sociology' in Hanson, A. H., Nossiter, T. and Rokkan, S. (eds), *Imagination and Precision in Political Analysis*, London: Faber and Faber.

Masterman, M. (1970), 'The nature of a paradigm' in Lakatos, I. and Musgrave, Alan E. (eds), *Criticism and the Growth of Knowledge*, Cambridge University Press.

Mulkay, M. J. (1970), 'Paradigms and cognitive norms', unpublished paper, Cambridge University.

Mullins, N. C. (1966), 'Social networks among biological scientists', Ph.D. thesis, Harvard University.

Mullins, N. C. (1972), 'The development of a scientific speciality: the phage group and the origins of molecular biology', *Minerva*, January, pp.51–82.

Popper, K. R. (1968a), 'Epistemology without a knowing subject' in van Rootselaar, M. and Staal, J. F. (eds), *Logic, Methodology and Philosophy of Science*, 3, Amsterdam: North Holland.

Popper, K. R. (1968b), 'On the theory of the objective mind', *Proceedings of the 14th International Congress of Philosophy*, pp.25–53, Vienna.

Ravetz, J. R. (1971), *Scientific Knowledge and Its Social Problems*, London: Oxford University Press.

Robertson, R. (1970), *The Sociological Interpretation of Religion*, Oxford: Blackwell.

Small, J. (1972), 'Prout's hypothesis: an historical/sociological study of its development in the nineteenth century with reference to models for scientific change', unpublished M.Sc. thesis, Manchester University, October.

Whitley, Richard D. (1972), 'Black boxism and the sociology of science' in Halmos, P. (ed.), *The Sociology of Science*, Keele University, Sociological Review Monograph No. 18, September.

5

Scientific knowledge and social control in science: the application of a cognitive theory of behaviour to the study of scientific behaviour*

Rolf Klima

Introduction: toward a sociology of scientific knowledge

One of the main topics of the current discussions about an appropriate sociological theory of scientific processes is the question of the relevance of the so-called cognitive aspects of scientists' activities. 'Cognitive factors' or 'cognitive variables' are used in this context as a comprehensive term denoting the nature of and differences between scientific ideas, their substantial meaning, logical consistency and coherence, the nature of scientific thinking, etc., in short, the whole of scientific knowledge both as the specific product of scientists' activities and as an important determinant of what scientists do, think and produce (cf. Whitley, 1972, p.61). The problem of how to take the structure and the development of scientific knowledge into account has attracted the attention of sociologists of science since philosophers of science as well as sociologists of knowledge have effectively questioned the prolificity of the hitherto prevailing 'Mertonian' version of the sociology of science. This had treated science 'not as a body of knowledge or set of investigatory techniques but as the organized social activity of men and women who are concerned with extending man's body of empirical knowledge through the use of these techniques' (Storer, 1966, p.3). The critics hold that this definition of the field has led to investigations of the social institutions and organiza-

* I am grateful to Heine von Alemann, Hartmut Lüdtke, Richard Münch, Werner Rammert, Ina S. Spiegel-Rösing, Helmut F. Spinner, Heinz Trusch, Ludger Viehoff and Hanns Wienold for comments on an earlier draft of this chapter.

tions of science, of scientists as members of a profession and of the communication processes within science while excluding 'any discussion of the subject matter of science' (Whitley, 1972, p.61). The development of scientific ideas is almost totally overlooked.

King (1971) and Martins (1972) have traced the neglect of the substance of scientific thought by the sociology of science back to the Mannheimian tradition of *Wissenssoziologie* which 'emphatically excluded logico-mathematical and natural-scientific knowledge from the purview of the "sociology of knowledge"' (Martins, 1972) because of the alleged immunity of scientific knowledge from social determination. This epistemologically based distinction has resulted in a bifurcation and disjunction of the 'sociology of knowledge' and the 'sociology of science': 'The sociology of science has not been deployed as a sociology of scientific knowledge and the sociology of knowledge has largely excluded science and scientific knowledge' (Martins, 1972). By accepting the view that scientific thought is scientific only in so far as it is governed by reason and logic, by the 'timeless' principles of scientific (i.e. logico-empirical) methodology and *not* by 'social factors', the Mertonian sociology of science could thus restrict its explanatory task to the study of the factors governing the social *behaviour* and the social institutions of 'scientists', defined as the people concerned with extending 'scientific' knowledge through the application of timeless principles of scientific rationality. Therefore, only 'the social influences that produce conformity to scientific norms and values' (Hagstrom, 1965, p.1), only the scientist's commitment to the particular goal of his profession, 'the extension of certified knowledge' (Merton, 1957, p.552), were taken as problematic. The scientific 'norms and values' themselves as well as the contents of the scientific ideas produced under their rule were taken as non-problematic for sociology, ceding the investigation of the conditions of their development to the historians and philosophers of science. The sociologists took the contribution of a scientist to the growth of knowledge into account chiefly by counting the number of his papers and/or citations without analysing their scientific content (cf. Lakatos, 1971, n.68; Musgrave, 1971, p.288).

Attempts to reintegrate the sociology of science into the sociology of knowledge and thus to problematize again the possible effects of 'social factors' on the *content* of scientific knowledge and its growth were stimulated greatly by Kuhn's *The Structure of Scientific Revolutions* (1962). His central hypothesis that 'normal' science is governed, not by a timeless, ahistorical and generally applicable canon of methodological rules leading to cumulative growth, but rather by specific traditions or 'paradigms' which tightly and with relative arbitrariness circumscribe the range of legitimate problems and methods of problem-solving, has opened the search for social factors conditional for the selection and acceptance of such 'paradigms'. The result is to reopen (at least for non-Marxist sociologists) the problematic of the 'social roots' of scientific thought.

The Kuhnian rehabilitation of the sociology of knowledge approach to the study of science bridges the gap between 'science as a particular sort of knowledge' (studied by the philosophy and history of science) and 'science as a particular sort of behaviour' (studied by the 'Mertonian' brand of sociology of science; cf. King, 1971, pp.3–4) through considering '*social factors*' (for example, authority) as an important set of independent variables which possibly influence the nature of the scientific ideas. Obviously, the reaction of the Popperian philosophy of science against Kuhn's alleged 'epistemological irrationalism', 'social-psychologism' and 'historical relativism' (Popper, 1970; Lakatos, 1970, pp.93, 178) has motivated other sociologists of science to look for a link between the two views of science by going the opposite way, namely by treating the nature of, and differences between, scientific ideas, 'epistemologically rational criteria', or '*cognitive factors*' as independent variables, too, which may influence the decisions of scientists (cf. Whitley, 1972, p.72).

Thus, at the moment it appears two orientations for the sociology of science are available: while the 'Kuhnians' attempt to explain the nature of scientific ideas, their production and acceptance with the help of 'social factors' like consensus, authority, power, stratification, etc., the 'Popperians' tend to emphasize 'cognitive factors' like certain criteria of epistemological rationality as determinants of scientific behaviour. These two perspectives will have to be theoretically reconciled. A theoretical approach is needed which allows the systematic study of the 'cognitive factors' as well as the 'social factors', both as independent variables and as dependent variables, and the processes which bring about the relationship between these factors within one systematic framework. In this sense, I take up the proposal to create 'a "translucid box" sociology of science which seeks to answer the questions: how do social and cognitive factors interact to produce knowledge and what effect do different forms of scientific knowledge have on society' (Whitley, 1972, pp.63f). Such a 'translucid box' sociology of science should specify that particular mechanism which transforms certain 'social' or 'cognitive' stimuli or 'inputs' to a scientist or a group of scientists into a specific 'output', i.e. behaviour on the part of the scientist(s), in particular the production of a specific kind of scientific information or knowledge.

My intention in this paper is to show that the existing psychological and social-psychological 'theories of cognitive consistency' and, more generally, modern approaches to the development of a cognitive theory of human behaviour could profitably be taken as the *starting point* for the development of a theory of scientific behaviour. For demonstrating the fruitfulness of such a strategy I shall refer primarily to Richard Münch's *Mentales System und Verhalten: Grundlagen einer allgemeinen Verhaltenstheorie* (1972).

Theories of cognitive consistency, cognitive behavioural theory, and the problems of the sociology of science

Since S. B. Feldman's (1966) and, later, R. P. Abelson's *et al.* sourcebook on *Theories of Cognitive Consistency* (1968), this term has become a customary family name for a set of theories which deal, roughly speaking, with the problem of what people do if they realize that there is a contradiction or 'inconsistency' between their beliefs, attitudes and/or behaviours or a 'conflict' between their different goals or needs. The exact, although varying, meaning of 'consistency' and 'inconsistency' only can be defined within the context of the different theories. In general, all the theories start with the assumption that people operate under a strong need for internal consistency or harmony among their beliefs, attitudes, etc. Since many of these theories refer to the tendency to conserve consistency among the affective and cognitive components of one's attitudes, assuming that an inconsistency leads to its change in order to re-establish harmony or balance, they have also been labelled 'theories of attitude change' (cf. Insko, 1967).

Since these theories have explicitly been formulated to give an account of cognitive processes, i.e. of the structure and functioning of human thought processes (see McGuire, 1968a), one can only be surprised that no one has systematically tried to apply them in the sociology of science. I have only found a single paper by M. Mulkay applying Festinger's notion of cognitive dissonance to a case of massive resistance by a scientific community against a 'deviant' scientific idea (see Mulkay, 1969, pp.37–9), and short remarks by the historian Mandelbaum in Abelson *et al.* (1968, pp.539–43) 'that scientific inquiry . . . can be looked upon as an attempt to reduce dissonance'.

Which specific properties make the consistency theories interesting for the sociologist of science, who is content neither with the existing functionalist and exchange theories of scientific behaviour which do not take the cognitive processes into account, nor with the philosophical theories of scientific progress which do not specify, at least not in a systematic way, the social-psychological conditions under which 'rational' cognitive behaviour, resulting in scientific progress, takes place? These important properties are, first, that they are *cognitive* theories, i.e. theories using cognitive or 'mental' variables in their explanations, thus allowing for the structure of knowledge-systems as determinants of people's behaviour, and second, that they are *disposition* theories, i.e. theories which try to specify the conditions under which a person will be motivated to behave in a specific (for example, in an 'epistemologically rational') way. The concept of 'cognitive inconsistency' can be viewed as a motivational construct (see Abelson *et al.*, 1968, pp.301ff) for the explanation of the mechanism through which certain stimuli or informational 'inputs' to a given cognitive system (including beliefs concerning the nature of concrete and abstract objects, cognitions concerning the needs, wants and goals of the individual, and evaluations) are transformed into a behaviour which results in the conservation or restoration of a state of harmony among the elements of that cognitive system, as defined by the rules of psycho-logic or whatever the definition of 'equilibrium' of a cognitive system may be within the context of one of the various consistency theories. Because scientists' activities can easily be conceptualized as a behaviour which aims at the collection of new information and its integration into a given knowledge-system either by interpreting the new 'data' in consistency with the given knowledge-system or by an appropriate reorganization of the given knowledge in consistency with the new information, it is obvious that the research tradition summarized under the name 'consistency theories' may have relevance to the analysis of scientific behaviour.

Thus, within the conceptual frame of consistency theory, the problem of 'scientific rationality' could be treated as the question under which conditions the assumed need for affective-cognitive consistency or consonance motivates the scientist to follow a research strategy which can be accepted as 'rational' according to the criteria of some philosophy of science.

If we accept the idea of the potential fruitfulness of theories of cognitive consistency for the sociology of science, we must, however, consider some problems which are obviously involved. First, is the conceptual heterogeneity of the various approaches which are, of course, not integrated simply through sharing a common family name. This heterogeneity is reflected already by the great variety of terms used for denoting the postulated point of equilibrium of a set of cognitive elements, namely 'consonance', 'balance', 'congruity', 'affective-logical', 'psycho-logical' and other kinds of 'consistency', not to speak of the rather different and generally vague meanings which are ascribed to them through the diverging axioms and hypotheses by which they are introduced. On the conceptual level, therefore, often completely different kinds of 'consistency' or 'inconsistency' are treated as identical (for example, logical contradictions between cognitions and incongruities between norms and actual behaviour), while factually identical consistencies-inconsistencies are sometimes divided into different categorical classes (cf. Münch, 1972, pp.3, 143ff). This confusion makes it difficult to compare these different approaches. Thus, for example, there has been a long-term disagreement between dissonance theory and a so-called reward-incentive-theory which are believed to lead to contradictory predictions (especially in relation to the famous 'Twenty-Dollar-Experiments' on the problem of forced compliance; cf. Aronson, 1968; Insko, 1967, pp.223–44), although an exact conceptual analysis of these two theories shows their complete identity and, therefore, also the identity of the predictions derivable from them (see Münch, 1972, pp.154–63). Also, that much of the empirical research performed to test various assumptions of consistency theory, in particular Festinger's hypothesis on selective exposure to inconsistent information, did not lead to corroborating results can probably be traced to insufficient conceptualization (cf. McGuire, 1968b; Sears, 1968; Münch, 1972, pp.167–73). In short, conceptual clarification of these approaches is overdue. A second problem, especially relevant for a theory for explaining scientific behaviour *per se*, and not only scientific thought, is the limitation of consistency theories to explanation of behaviour related to cognitive or attitude change by individuals, thus excluding social behaviour in general.

These considerations motivate the demand for a general behavioural theory which could be applied to both the problems which the consistency theories attempt to solve and those which the reward-punishment or stimulus-response theories, constituting the basis of approaches like exchange-theory, attempt to solve.

R. Münch's 'incongruity theory' is important because it tries to outline the fundamentals of such a cognitive behavioural theory which, among other possible applications, can also be used for a clarification and at least partial integration of the various theories of cognitive consistency and for the solution of some central problems in this area. Because of the generality, conceptual clarity and relative simplicity of Münch's version of a general cognitive theory of human behaviour it seems worthwhile to attempt formulating, in terms of his approach, the elements of a future theory of scientific behaviour. In the following section we shall, in an abbreviated form, introduce the most important concepts and assumptions of the incongruity theory.

Basic concepts and statements of incongruity theory

(a) *Introductory remarks*

Any behaviour can be seen as being caused by two types of conditions: (a) the conditions for an individual to be disposed to a specific behaviour and (b) the conditions which determine the objective possibilities of an individual to behave in a specific way (see Münch, 1972, pp. 23 ff). If one accepts this view, the existence of a disposition is only a necessary, and not a sufficient, condition for the behaviour the person is disposed to. In addition, the person must have the possibility to act in this way. Besides the individual's aptitudes or personal abilities, the relation of the individual to his environment is the most important set of conditions on which the possibility of a particular behaviour depends: the environment determines the degree of freedom to behave in accordance with one's goals. Of course, the central problem of most sociological theories of the macro-level type is the specification of the restrictions and opportunities set by a particular social environment for the possibilities of individuals or classes of individuals to act in a specific way. In contrast to such approaches, incongruity theory is a disposition theory which takes the influence of the environment into account only in so far as it determines the individual's cognitions. Therefore, also a 'theory of scientific behaviour' derived from incongruity theory will be a disposition theory, stating only the conditions for scientists to be disposed to a specific scientific behaviour. A complete explanation of scientists' activities, however, would also require theoretical propositions concerning the structure of possibilities or the 'structural constraints' which framed the scientists' activities. It is not intended, in this chapter, to develop such propositions.

Incongruity theory, as a disposition theory, states that any behaviour to which a person is disposed aims at the reduction of a perceived discrepance between the goals of the person and the degree of realization of these goals. The process through which such a disposition is produced is conceptualized as a decision-making process in which the person 'draws conclusions' from the information he has about the situation and his information about the means which are necessary or sufficient for reaching his goals in a given situation. Thus, in principle, the theory gives a hypothetical description of the 'black-box' mechanism which transforms certain informational inputs to the 'mental system' of a person into dispositions to act in a specific way, which then become—together with the opportunity to act in this way—causal conditions of a particular behaviour.

(b) *Predications*

The basic units of the mental system are predications. Münch suggests that all 'mental predicates', i.e. attributes which can be used for the description of the mental state of a person, like 'believing', 'hoping', 'wishing', 'demanding', 'knowing', 'being motivated', etc., denote the relationship of a person to a certain predication, namely a possessive relationship. A predication is defined as the content or the meaning of a proposition (a proposition being logically defined as the attachment of a predicate-sign to an argument). Thus all sentences using mental terms, like 'believing', 'hoping', etc., can be transformed into sentences expressing the possessive relationship to a predication or a class of predications.[1] So, for example, the sentences:

'Lakatos knows the methodology of naive falsificationism'
'Lakatos believes that the methodology of scientific research programmes gives epistemologically rational criteria for scientists' activities'
'Lakatos demands the rational reconstruction of the history of science'

can be transformed into:

[1] Throughout this chapter when we want to refer to a particular predication we shall use the name of this predication and diagonal slashes (/) before and after it. The proposition which expresses a predication can be used as its name.

'Lakatos possesses true cognitions about the methodology of naive falsificationism'
'Lakatos possesses the belief/The methodology of scientific research programmes gives epistemologically rational criteria for scientists' activities/'
'Lakatos possesses the demand/The history of science should be rationally reconstructed/'.

Now, after introducing the 'possession of a predication' as the basic element of all mental attributes of a person, the whole set of predications which are in the possession of a person can be defined as his mental system (Münch, 1972, p.42). It may be noted that here 'predication' is called what in most theories of cognitive consistency (e.g. in Festinger's dissonance theory) is called 'cognition' or 'cognitive element'.

(c) *Cognitions and standards*

The next important step is a basic partition of the predications, i.e. the elements of a mental system, into two subsets. The criterion of this partition is the modality of the predications. Predications of the cognitive modus, i.e. predications which express an assertion or description of something, are called cognitions; predications of the normative modus, i.e. predications which express that something is prescribed, demanded, required, necessary, needed, prohibited, etc., are called standards. This discrimination between 'cognitions' and 'standards' is perhaps the most decisive conceptual difference between consistency theory and incongruity theory which already clarifies many of the difficulties of consistency theory. That standards and cognitions are really two different types of predications, and not, for example, standards a specific type of cognitions or vice versa, can easily be shown by trying to transform a standard like 'I should not lie' into a cognition, for example: 'There is a standard which prohibits me to lie.' Such a transformation is impossible because having a standard or norm is not equivalent to having the knowledge that such a standard or norm exists. Thus, also a 'dissonance' between two cognitions or between two standards and a 'dissonance' between a cognition and a standard cannot be treated as theoretically equivalent things as dissonance theory does. One has to make a difference between (logical) 'inconsistency' and 'incongruity'.

(d) *Inconsistencies*

The relation of (logical) inconsistency can only exist between two predications of the same modus, i.e. between two cognitions or between two standards, but not between a standard and a cognition. A logical inconsistency between two predications exists if and only if one of them implies the negation of the other one, symbolically: /p and non-p/ or—using the sign '!' for designating a standard—/!p and !non-p/. There is no such logical contradiction between a standard demanding a particular state of affairs and a cognition asserting the non-existence of this state of affairs, symbolically: /!p and non-p/. !p demands, but does not imply non-non-p (or p), as non-p does not imply !non-p. For incongruity theory, the presence of an inconsistency alone has no behavioural consequences; it is not assumed that individuals have a general need for logical consistency among their beliefs or norms.

(e) *Incongruities*

The just mentioned relation between a standard of a person demanding a particular state of affairs and a cognition negating the existence of this state—/!p and non-p/ or /!non-p and p/—is called an 'incongruity'. Thus, the (combined) predication /I should not lie, but I am a liar/ would be an incongruent predication; there would be incongruity within the mental system of the person possessing this predication.

The cognition of such an incongruity is suggested to be the necessary and sufficient condition of any behaviour for which the possibility is given. It constitutes the disposition to a behaviour which aims at the reduction of the incongruity and its reversal into congruity. This assumption is the basic axiom of Münch's incongruity theory. It is an axiom because this assumption itself cannot be explained within the context of this theory: the explanation that the tendency towards congruity is itself caused by an incongruity between a standard or motive of the type /I should not experience incongruity/ and the cognition that there is such an incongruity, would lead to circularity. Therefore, the incongruity between a standard and a cognition is viewed as the 'most general necessary and sufficient condition preceding any behaviour. Therefore there cannot also be individual differences in relation to the striving for congruity' (Münch, 1972, p.96).

There can, on the other hand, be individual differences in relation to the motivation or disposition to reduce logical inconsistencies within their own system of cognitions or norms. These differences may range from extreme intolerance to extreme tolerance of inconsistency, depending on the magnitude of incongruities produced by a given type of inconsistency. In principle, a person is disposed to reduce a given inconsistency if and only if he possesses, besides the cognition of the particular inconsistency within his mental system, a standard prohibiting an inconsistency of this particular type or of inconsistencies in general. Only under this condition, will there be an incongruity constituting the disposition for an inconsistency-reducing behaviour. Under certain conditions even an increase of inconsistency may serve to reduce incongruities, as will be discussed later. This may explain many of the contradictions among, and experimental falsifications of, hypotheses derived from consistency theories; it must be assumed that, in cases where an expected consistency-striving could not be observed, the postulated consistency standard did not rank sufficiently high or was even absent (cf. Münch, 1972, pp.137–9, 143–73).

(f) *The theory*

As we said, the existence of incongruities within a mental system is the condition for any behaviour. For predicting which particular behaviour will take place, specification of the basic assumptions is necessary. It will be necessary to state the conditions determining which of the many possible activities will be chosen to reduce an incongruity, and which incongruity will be chosen if there are several within the mental system. The strategy which Münch uses to answer this question is to specify the behaviour to be selected by the individual as that one which reduces the 'highest' incongruity to the greatest extent and leads to the lowest 'secondary' incongruities. This presupposes the existence of comparative predications within the mental system which allow a rank ordering of different phenomena to which predications refer and, therefore, also of different incongruities (see Münch, 1972, pp.80–9). Now, the theory can be formulated as follows (Münch, 1972, p.99):

> If and only if a person possesses cognitions about his own incongruities, then he selects that behaviour which reduces the incongruities of the highest rank to the greatest degree and which produces incongruities of the lowest rank to the least degree.

This formulation takes into account that most activities may not only reduce incongruities, but also produce new incongruities or have certain 'costs'. It postulates that exactly that behaviour is selected which produces the highest 'net-congruity'.

As mentioned above, the mental behaviour leading to a disposition to a specific behaviour can be analysed as a decision-making process. Münch (1972, p.101) distinguishes three logical, not necessarily chronological, steps of this process. First, the selection of the incongruity to be reduced from the set of incongruities existing within the mental system at a given moment. Second, the selection of the behaviour for the reduction of the incongruity already selected and, third, the detailed implementation of the behaviour selected at the previous step. It will be necessary to analyse this process in greater detail before we can apply the incongruity theory to more complex phenomena like the behaviour of scientists in actual social settings. Principally, this analysis will give us a general model of the 'situational logic' (cf. Popper, 1966, II, pp.89–99) which can be used as a heuristic for the reconstruction of any decision-making process. To construct a sociological theory of scientific behaviour it will, then, be necessary to introduce certain simplifying assumptions which specify the general model with regard to the historically concrete conditions of the situation in which the scientists act.

(g) *Decision step I: selection of the primary incongruity to be reduced*

Incongruity theory postulates that incongruities can be rank-ordered. As far as mere classificatory incongruities are concerned—i.e. incongruities involving cognitions and standards of the either-or-type: either I (should) have predicate x or not—the rank of an incongruity depends solely on the contribution of its reduction to the production or reduction of other incongruities. As far as metric incongruities are concerned—i.e. incongruities involving cognitions and standards of the more-or-less-type: I (should) have predicate x to the degree d—the rank of an incongruity depends also on its magnitude. The magnitude of a metric incongruity is defined (Münch, 1972, pp.92, 114ff) as the degree of the deviation of the perceived reality from the state of affairs demanded by the standard. It is further postulated by incongruity theory that, if in a particular situation there are several incongruities existing in the mental system of an individual, he will decide to reduce the incongruity of the highest rank.

Thus, as the first step of the decision-making process, he will have to decide which of the several incongruities should be considered as ranking highest in the given situation. The individual will, therefore, compare these incongruities with regard to differences of their magnitude and of the effects which their retention or reduction will have on the production or reduction of other ('secondary') incongruities. In principle, this comparison process can be understood as a mental activity which, after producing comparative predications of the form /incongruity i_1 is greater than incongruity i_2 and incongruity i_2 is greater than incongruity i_3 . . ./, conditional predications of the form / reduction r_1 leads to (secondary) incongruity $(s)\ i'$ and reduction r_2 leads to (secondary) incongruity $(s)i''$ . . ./, comes to conclusions similar to /i_1 ranks higher than i_2 and therefore r_1 must be preferred to r_2/, etc. In addition, the rank-ordering is influenced by the degree of security with which the individual accepts the truth of the conditional predications serving as the basis of comparison. If the conditional predication /r_1 leads to $(s)i'$/ seems much more insecure than the conditional /r_2 leads to $(s)i''$/ and $(s)i'$ is ranking only slightly higher than $(s)i''$, then—*ceteris paribus*—r_2 will be preferred to r_1. This is, of course, the well-known problem of 'expected utility'.[1]

(h) *Decision step II: selection of the behaviour for the incongruity reduction*

Having selected the incongruity with the highest rank in a given situation it is now necessary to decide which form of behaviour should be selected to reduce the incongruity. Münch (1972, p.117) enumerates the following types of behaviour which can be used for reducing an incongruity:

(A) Alteration of the conditional cognitions about the contribution of the reduction of an incongruity to the reduction of a secondary incongruity.

This leads to a reduction of the rank of the incongruity through alteration of the rank of the involved standards. A behaviour seems less demanded if the expected positive effects appear less probable.

[1] In order to make our following discussion of the elements of a sociological theory of scientific behaviour not too complicated, we shall neglect in this discussion the role of the magnitude of an incongruity and the role of the certitude of the belief in the predications involved in an incongruity for the decision-maker.

(B) Alteration of the cognitions about the deviation of reality from the standard through:
  (a) deception,
  (b) dissolution of a deception by additional information,
  (c) alteration of reality itself.

Which behaviour will be selected depends completely on the rank and magnitude of the incongruities produced and reduced in this way. In any case, a behaviour that is expected to lead to higher incongruities than the incongruity to be reduced will never be chosen.

(i) *Decision step III: implementation of the selected behaviour*

The final step in the decision-making process is mainly important for the individual who has already decided in favour of an incongruity-reduction by changing the reality. Here, a plan or strategy has to be developed and/or if at hand executed.[1] If a person wants to reduce an incongruity by changing reality and does not have a plan, one says that he has to solve a problem. Also, this problem-solving behaviour is guided by decision-making processes following the same pattern as in step I and II.

Elements of a 'translucid box' theory of social control in science

The model of the mental or 'cognitive' process which leads to a particular disposition can now be used for specifying the questions which will have to be answered if we are to formulate a 'translucid box' theory of social interaction and control in science. These questions are:

(a) What are the primary incongruities whose reduction could be considered constituting the main activity of scientists?

(b) What are the secondary incongruities which determine the reduction behaviour of scientists with regard to their primary incongruities?

(c) Given certain essential primary and secondary incongruities, how does the scientists' behaviour aimed at the reduction of these incongruities relate the scientists to their social environment?

We shall introduce the assumptions answering these questions step by step.

[1] For a behavioural theory which interprets behaviour as the construction and execution of 'plans' see Miller, Galanter and Pribram, 1960.

(a) *Logical inconsistencies as the root of primary incongruities*

'Primary incongruities' are those incongruities present in the mental system of a person which, in a given situation, constitute the 'starting point' of his decision-making process which, in turn, results in a disposition to a particular behaviour (Münch, 1972, p.104, n.71). They are those incongruities which directly refer to the given situation, i.e. involving cognitions and standards referring to the given situation, as defined by the individual, and the behavioural alternatives which are open to him in his particular situation.

If we analyse these primary incongruities, the reduction of which could be considered as the main activity of scientists, this amounts to the question: what constitutes the 'typical' decision-problems of a scientist *qua* scientist? In other words, what are the primary problems the solution of which defines the role of the scientist? If we consider all behaviour as a reaction to a particular incongruity between standards and cognitions: what are the specific incongruities to be resolved by 'scientific behaviour'? We need, so to say, an 'analytical definition' (see Hempel, 1952, p.8) of those aspects of the behaviour of the people who occupy scientific roles and positions which must be considered as genuinely 'scientific' and thus as the explanandum of a sociological theory of science. In this sense, indeed, we need a 'demarcation criterion' (cf. Lakatos, 1971, p.106ff); of course, the sociology of science does not deal with the behaviour of 'scientists' as husbands, politicians, industrial managers, journalists, artists, etc. but only with the behaviour of scientists as scientists.

In connection with the discussion of the potential relevance of the theories of cognitive consistency for the sociology of science, we have already attempted to conceptualize the activities of scientists as a behaviour which is primarily directed towards the production of consistency within the system of cultural knowledge. We now assert that any scientific research can be interpreted as an activity which aims at the resolution of inconsistencies between predications. There are different epistemological theories concerning the questions (a) under which conditions an inconsistency can be viewed as being reduced and (b) by which method an existing inconsistency should be reduced. The first is a problem of the logical analysis of the structure of knowledge-systems, while the second refers to the problem which consistency standards should guide the inconsistency reduction. Depending on the answers given to these questions, there are different conceptions of 'scientific rationality' and often also different opinions on the so-called 'demarcation criterion' used to separate 'science' from 'non-science' (cf. Lakatos, 1971, pp.49–51). But it seems reasonable to assume that scientific activity must at least lead to the reduction of inconsistencies. The sociology of science should not only account for what a philosophy of science defines as 'good' science. It may be objected that scientists often do exactly the opposite of reducing inconsistencies, namely producing inconsistencies—for example, by collecting data or formulating theories which contradict established theories. This can, indeed, be an important aspect of scientific activity but it should be noted that the goal to reduce existing inconsistencies does not necessarily imply a standard to avoid possibly inconsistent information.

If our analytical definition of 'scientific behaviour' as 'inconsistency reduction' is valid, then we are justified in considering incongruities involving cognitions and standards which refer to the existence or non-existence of logical inconsistencies within scientific knowledge as the primary incongruities which determine the decision behaviour of scientists. In other words, assuming that there are several incongruities in the mental system of the scientist, we postulate that—analytically stated—at the 'first step' of his decision-making he will usually choose to reduce those incongruities referring to problems of logical inconsistency. This presupposes that incongruities concerning logical inconsistencies are usually considered as those which occupy the highest rank among the incongruities which may be realized by a scientist in his professional role. Cognitions and standards referring to existing inconsistencies within the body of knowledge for which a scientist is considered to be responsible constitute the frame of reference within which he works and which primarily determines his decisions.

A theory for the explanation of the treatment of logical inconsistencies by scientists should account both for the disposition of scientists to resolve given inconsistencies and for the possibility that they will tolerate or even actively seek inconsistent information in order to reduce higher ranking inconsistencies. Thus, as a first basic theorem for the explanation of scientific behaviour, we may assert:

(1) If and only if a scientist possesses cognitions about an incongruity between his standards and his cognitions referring to the existence or non-existence of logical inconsistencies within scientific knowledge, then he will reduce those inconsistencies which produce incongruities of the highest rank to the greatest degree and reduce incongruities of the lowest rank to the least degree, and he tolerates

and seeks those inconsistencies which reduce incongruities of the highest rank to the greatest degree and produce incongruities of the lowest rank to the least degree.

Following incongruity theory, the next question is which (secondary) incongruities can be produced or reduced by logical inconsistencies. In other words, which standards are threatened by the existence or non-existence of inconsistencies? As mentioned, incongruity theory does not assume the existence of absolute or 'categorical' consistency standards. Therefore, the reasons for the reception and operation of consistency standards within the mental systems of persons and the reasons for their social institutionalization can themselves be treated as a problem for theory and empirical research. As a matter of fact, the assumption that people possess a categorical imperative to avoid inconsistencies under every condition would *a priori* exclude the possibility of opinion and attitude change; there is even a contradiction implied in traditional dissonance theory in so far as this theory, at the same time, postulates that people generally avoid inconsistent information and that cognitive and attitude change is a result of the reduction of previously experienced inconsistency (cf. Münch, 1972, p.168). Thus, it seems to be more reasonable to assume that the consistency motive usually functions as a sub-standard in the service of other standards or that, the appearance of logical inconsistencies in the mental system is usually undesirable, not because people 'just don't like inconsistency', but because the existence of inconsistencies may have undesirable consequences of different kinds. This assumption has far-reaching implications for the prediction of the behaviour selected by people experiencing or expecting inconsistency; which behaviour they will select will depend on the type of standards referring to a given or expected inconsistency, and the relation of these standards with other standards and cognitions.

(b) *Secondary incongruities which determine the treatment of logical inconsistencies*

Münch (1972, pp.163ff) has speculated about different types of standards which may be incongruent with the toleration of logical inconsistencies between cognitions or standards. According to him, one of them may be the standard demanding an orientation toward the environment which helps to solve problems in the best possible way. If one believes only true cognitions are suitable for such an orientation, then from this standard and this belief follows the standard: /I should not maintain false cognitions/. If, in addition, one has the cognition /If I have two inconsistent cognitions, at least one of them is false/, then it can be deduced: /If there are inconsistent cognitions, then the false cognition(s) should be eliminated/. Thus, in this case, the consistency motive functions as a sub-standard in the service of the standard/! problem solving/. Under this condition a strategy to reduce inconsistencies by misperceiving the cognized reality or by avoiding inconsistent, but possibly true cognitions, which is usually assumed by the traditional consistency theories, would, of course, be inapplicable.

But a consistency standard may also be inferable from standards prohibiting the alteration of established dogmas, a deviation from an established cognitive consensus, a change of beliefs once publicly professed by oneself, the acceptance of ideas which threaten the legitimation of power, etc. For example, a consistency standard of this type may be inferred in the following way:

/My knowledge should be useful for the emancipation of underprivileged people/

/Only cognitions which do not contradict Marxism are useful for the emancipation of underprivileged people/

/If there is an inconsistency between Marxism and another cognition, then this other cognition should be eliminated/.

In this case not only a 'conservative' strategy for the reduction of inconsistencies already apparent would be demanded, but also the avoidance of information inconsistent with the belief-system to be conserved: /I should avoid informations which contradict Marxism/. If, however, the consistency motive is a sub-standard of the standard /!growth of true knowledge/, then the admission of inconsistent informations would surely create incongruities, too, but these incongruities would occupy a lower rank than the incongruity which would be produced by the avoidance of inconsistent, but possibly true, cognitions.

Most philosophers of science will probably agree that the only consistency standard which is compatible with 'scientific rationality' is the one inferred from the standard /!growth of true knowledge/. As a matter of fact, the epistemological theories of scientific rationality must be considered as more or less elaborate strategies for the treatment of logical inconsistency, most of them presupposing the validity and superiority of the standard /!growth of true knowledge/ (cf. Münch, 1972, pp.177ff). A sociological theory of science, however, should make no *a priori* assumptions about the 'rationality' followed by scientists in their treatment of inconsistencies; instead, it should state the *social* conditions under which scientists will probably follow one or the other strategy or perceive and interpret the 'inner logic' of a given knowledge system in one or another way (cf. Whitley, 1972, p.84f). Therefore, in such a sociological theory, the various possible consistency standards should themselves be related to standards (and, thus, to secondary incongruities) which refer to the social environment in which the scientific behaviour takes place.

At this point of our discussion, again we need a 'demarcation criterion'. We have to decide what type of social setting our theory should include. Scientists work in many settings and under different conditions, but only a general theory of action—like incongruity theory—can refer to all possible conditions. In accordance with the exchange theories of social control in science, presented in the writings of Hagstrom (1965), Storer (1966), Scherhorn (1969) and others, we shall, for now, restrict our explanatory goals to the explanation of scientists' activities under 'market' conditions, where the scientists exchange, among each other, scientific contributions for professional reputation.

As will be remembered, the reputation-exchange theory asserts that other things being equal, the scientist chooses that behaviour (e.g. decides to work in that problem area) which will lead to maximum professional recognition and reputation. The explanations given by different authors for the importance of this reward-type for the scientist diverge. Thus, Storer assumes the necessity of recognition as the condition for the satisfaction of a basic creativity need; Hagstrom, on the other hand, seems to assume that the need for recognition can be traced back to socialization experiences (cf. Hagstrom, 1965, p.9). All exchange theorists assume, however, that the desire to gain professional recognition is, if not a consequence of, at least strongly reinforced by the structurally guaranteed dependence of the professional scientist working in basic research on positive evaluations of his work by his colleagues as a precondition for material success, a career, research funds, etc. (see also Collins, 1968). But be that as it may, for the moment it seems a relatively well established assumption that scientists who are not primarily dependent on non-scientists are strongly motivated by a wish to gain reputation as a scientist. Therefore, we shall assume that this wish creates the most important secondary incongruities which the scientist takes into account in his decisions.

Using this assumption, we can now give a more detailed specification of the behaviour to which a scientist will be disposed if he realizes or expects logical inconsistencies within his scientific knowledge: he will be disposed to that behaviour with regard to these inconsistencies which, other things being equal, promises to lead to the greatest increase of his professional reputation. Through this assumption we can hope to relate propositions concerning the structure of scientific knowledge (its logical consistency and inconsistency) to the exchange theory of social control in science, and thus, to transform this theory into a 'translucid box' theory.

The influence of the (secondary) incongruities created by the scientist's wish to gain professional recognition may, within the systematic framework of incongruity theory, be stated as follows:

(2) If and only if a scientist possesses cognitions about an incongruity between his need for professional reputation and his actual level of professional reputation, then he will choose activities for increasing his professional reputation which reduce his highest-ranking incongruities to the greatest degree and which evoke his lowest-ranking incongruities to the least degree.

If we now use this theorem 2 for specifying our theorem 1 we may come to the following theorem:

(3) If and only if a scientist possesses cognitions about an incongruity between standards and cognitions referring to the existence or non-existence of logical inconsistencies within scientific knowledge, then the following holds true:
if and only if the scientist possesses cognitions about an incongruity between his need for professional reputation and his actual level of professional reputation, then he will reduce those inconsistencies whose reduction increases his professional reputation to the greatest degree, and he will tolerate and seek those inconsistencies the toleration and acceptance of which increases his professional reputation to the greatest degree.

Through this assumption the decision of the scientist whether to reduce or to tolerate inconsistencies in accordance with one or the other, scientifically 'rational' and 'progressive' or 'conservative', consistency strategy is made dependent on the increase of his professional reputation which he can gain by alternative decisions. Thus, it will be important to know through which strategy a scientist can increase his professional reputation in a scientific community which is organized as a 'scientific market' (cf. Scherhorn, 1969). How does his treatment of inconsistencies relate him to his relevant audience of professional colleagues?

(c) *The treatment of inconsistencies and the operation of social control in science*

We shall assume that the satisfaction of the desire for reputation will mainly depend on the degree to which the scientist will be able to reduce incongruities on part of his immediate colleague group, and that these incongruities are related, in the same way as his own primary incongruities, to the existence of inconsistencies within the system of scientific knowledge which constitutes the common universe of discourse of this 'invisible college'. Therefore, the scientist will be involved in a power-relationship as defined by Münch's following proposition (1972, p.26):

> The higher the rank and the greater the magnitude of y's incongruities of which y believes that x will be able to reduce or evoke them, compared to the alternative possibilities of incongruity reduction as perceived by y, the more y conforms to the perceived will of x in his decisions.

In traditional reward-punishment-terminology this means: y will be disposed to do what x wants if y has made the experience (cognition) that x can reward (reduce incongruities) or punish (evoke incongruities) him. In principle, this is the kind of social power which is the instrument of social control in 'markets'. The propositions of any exchange theory are based exactly on this assumption.[1]

For our purpose, this general power theory has to be specified for the following conditions: the incongruities, of which y (=the scientist) believes that x (=his colleagues) will be able to reduce or evoke them, are specified as incongruities related to his lack of reputation, while the 'perceived will' of x is specified as the wish to have certain logical inconsistencies in the common knowledge treated in a particular way. As a consequence, the scientist will consider the degree with which the

[1] In this case x has power over y because of y's cognitions about x's possibilities to reduce y's incongruities: these cognitions influence y's *dispositions*. A completely different type of power-relationship between x and y is given if x can influence y's *possibilities* to behave in a particular way, i.e. if x can restrict y's freedom. While the conditions for a power-relationship of the first type can be stated completely in terms of incongruity theory as a disposition theory, the explanation of a person's power to restrict the possibilities of another person makes a different theory necessary (see Münch, 1972, pp.25–6).

selection of a particular inconsistency-treatment will contribute also to the reduction of incongruities related to inconsistencies as perceived by his colleagues. This power-relationship can be expressed as follows:

(4) The greater the rank and the magnitude of a scientist's incongruities related to his need for reputation which can be reduced by his colleagues, compared to his alternative possibilities to gain reputation, the more the scientist will tend to treat inconsistencies within the knowledge of his field in a way which he believes is congruent with his colleagues' consistency standards and, therefore, induces his colleagues to supply him with reputation.

To make this power-relationship reciprocal, of course, also the conditions must be stated which account for the colleagues' willingness to provide a scientist with reputation. This is usually forgotten by exchange theorists in the sociology of science (G. Scherhorn is the only exception). We suggested already that this condition will be the degree to which the scientist reduces important incongruities of his colleagues, and that these incongruities will also primarily be related to the existence or non-existence of inconsistencies within knowledge:

(5) The greater the rank and the magnitude of a scientist's incongruities concerning the existence or non-existence of inconsistencies within his scientific knowledge which the scientist believes one of his colleagues will be able to reduce or evoke, compared to the alternative possibilities of reducing or evoking these incongruities (for example, by research performed by himself), the more this scientist is ready to provide this colleague with reputation.

It may be emphasized that in this 'translucid box' version of an exchange theory of social control in science it is not 'the social institution of science' (whatever this may be) which uses the allocation of recognition 'to advance knowledge' (cf. Whitley, 1972, p.71) but, rather, the individual scientists working in basic research who, in their role as 'consumers' of scientific information, use the allocation of recognition to advance the reduction of their own incongruities, the solution of problems which they themselves 'have'! Scientists, as demanders in scientific markets, reward colleagues who supply them with those problem-solutions which, according to their own standards, seem to be useful for the advancement of their own work for which they are responsible (cf. Scherhorn, 1969). F. Reif (1965, p.135) has stated this very clearly: 'The scientist is not different from others in his desire to be successful, but his definition of "success" has some distinctive features. The work of the pure scientist is abstract; it consists essentially only in gathering new data and formulating new concepts. To constitute scientific knowledge, these must be verifiable by other scientists and usable by them as the basis for further exploration', and, as we may add, for further exploitation. As a matter of fact, if only the recognition which a scientist may earn would reward him, but not the scientific information which he receives from his colleagues, there would be no 'exchange' at all.

(d) *Progressive and conservative strategies of inconsistency reduction*

If now we should wish to make predictions about the scientific behaviour of scientists it would be necessary to collect information about the standards and cognitions held by the group of their scientific colleagues with regard to the preferable treatment of inconsistency. It will be important to know, in particular, if this group holds standards which could lead to high-ranking incongruities when either a conservative or a progressive strategy of inconsistency reduction is selected by the scientists under investigation.

So far we have only discussed the decision behaviour of scientists at the first step of Münch's three-step-model of decision-making, when it has to be decided which incongruity to reduce. But all we said about the factors determining the decisions at this step, of course, holds true for the second and third steps, where the behaviour to reduce the selected incongruity has to be chosen, and the chosen behaviour to be implemented. Therefore, we can be very brief in discussing what will happen after a scientist has already decided to reduce an incongruity concerning given inconsistencies. In accordance with the catalogue of possibilities given above he has initially to decide:

(A) whether to alter his cognitions about the expected contribution of the reduction of the incongruity between his consistency standards and the inconsistencies in question to the reduction of secondary incongruities; this would give him the opportunity of lowering the rank of the involved consistency standards.

For example, the scientist could come to the conviction that he would not gain much reputation if he found a solution for the problems posed by certain experimental anomalies because these do not affect theoretical beliefs which are relevant for his colleagues.

If the incongruity cannot be reduced in this way, then he may:

(B) change his cognitions about the involved inconsistencies in one of the following ways:

(a) misinterpret the cognized facts which are involved in the inconsistencies so that the inconsistencies disappear:
(b) resolve the inconsistencies by the discovery of additional evidence showing that, as a matter of fact, the cognition of inconsistencies was erroneous;
(c) resolve the inconsistencies through alteration of one or more of the involved inconsistent predications.

A progressive strategy of inconsistency reduction would imply, according to many philosophers of science, that inconsistencies which are not produced by empirical or logical errors are reduced by formulating theories of greater empirical content. This, of course, presupposes that the scientist in question is ready to change his theoretical convictions. But the readiness to behave in this way may suffer if such a behaviour would be incongruent with his need to gain recognition from his colleagues. This may be the case, for example, if the established consensus serves as a legitimation for a given distribution of power and influence within a scientific community or if the belief in the truth of the established theories is so strongly supported by the knowledge accumulated within a given research tradition that a change of these theories would lead to more and greater inconsistencies (and incongruities produced by them) than a neglect or elimination of the cognitions which are inconsistent with the established belief.

The conservative strategy resulting from such conditions would demand a resolution of appearing inconsistencies between existing theories and new information ('anomalies') solely in favour of the old theories. In this way, perhaps, the role of Kuhn's 'normal scientist' can be understood. A 'revolutionary scientist', on the other hand, may be successful if the number and the magnitude of the inconsistencies already contained in the established knowledge of a scientific group has become so great that approaches for restoring consistency by means of a completely new and different 'paradigm' may appear as promising. In such a situation there may, indeed, be a strong demand for new theoretical outlooks which may stimulate a correspondingly intensive competition among scientists to supply their scientific community with the most satisfying paradigm (cf. Scherhorn, 1969, pp.76f). In a 'translucid box' theory of social control in science the possibility that a scientist may decide 'how to tackle a problem on the basis of epistemologically rational criteria' (Whitley, 1972, p.72) is not 'ruled out by the emphasis on social rewards and social exchange', because such a theory accounts for the conditions that these criteria or other standards may determine the dispositions of scientists both in their role as producers of scientific contributions and in their role as demanders and rewarders of innovations.

Concluding remarks

In this chapter I have tried to demonstrate that the existing cognitive theories of human behaviour may be useful for the explanation of scientific processes. Such an approach makes it possible, first, to consider the structure of scientific knowledge, as represented in the cognitions of scientists, as an important independent variable which influences the productive behaviour of scientists and, second, to conceive of scientific ideas as the outcome of a particular behaviour with which the scientist reacts to particular circumstances of his cultural and social environment. Both have been neglected by the traditional sociology of science. If we apply assumptions derived from current cognitive theories of human behaviour to the problems of the development of scientific knowledge, then the 'black box', within which certain stimuli from the cultural and social environment of scientists are transformed into certain scientific products, may become a 'translucid box'.

Further work along the lines of this theoretical framework should concentrate, in the first place, on the development of satisfactory observational theories stating which initial conditions should be observable if the theory is to be empirically tested (cf. Lakatos, 1971, pp.132ff). If the explanations inferred from incongruity theory are not to be tautological, it must be possible to decide whether a person possesses certain incongruities without referring to the behaviour which is suggested to be caused by the disposition to reduce the incongruities in question. With regard to the explanation of scientific behaviour, this implies the need to have measurement techniques through which the logical structure of scientific knowledge and the standards referring to this knowledge as represented in the mental systems of scientists can be assessed. Second, we must try to construct additional assumptions to account for the influence on scientific behaviour exerted by other social environments than the scientific 'markets' of basic research such as the organizations in which scientists work and the political and economic agencies which provide the material and social support for scientific activities. It seems important, in particular, to study how the scientist reaches a certain research decision if he depends for the reduction of his incongruities on various 'relevant audiences', and if these audiences themselves apply inconsistent standards with regard to what they demand from the scientist (cf. my own analysis of German sociology; Klima, 1972). Third, because we can explain a person's behaviour only if we know, in addition to the conditions of the dispositions of this person, the conditions of the possibilities of this person, we must formulate theoretical propositions about the conditions which account for a scientist's possibilities of reaching his goals in a given situation. The personal abilities of the scientist, his access to resources and various organizational constraints are probably the most important factors which have to be considered as conditions of the possibilities of scientific activities.

References

Abelson, R. P., McGuire, W. J., Newcomb, T. M., Rosenberg, M. J. and Tannenbaum, P. H. (eds), (1968), *Theories of Cognitive Consistency, a sourcebook*, Chicago: Rand McNally.

Aronson, Elliot (1968), 'Dissonance theory: progress and problems' in Abelson, R. P. *et al.*, op. cit., pp.5–27.

Collins, Randall (1968), 'Competition and social control in science: an essay in theory-construction', *Sociology of Education*, 41, 2, pp.123–40.

Feldman, Shel B. (ed.) (1966), *Cognitive Consistency: motivational antecedents and behavioural consequents*, New York: Academic Press.

Hagstrom, W. O. (1965), *The Scientific Community*, New York: Basic Books.

Hempel, Carl G. (1952), *Fundamentals of Concept Formation in Empirical Science*, Chicago: International Encyclopaedia of Unified Science, I, 7.

Insko, Chester A. (1967), *Theories of Attitude Change*, New York: Appleton-Century-Crofts.

King, M. D. (1971), 'Reason, tradition and the progressiveness of science', *History and Theory*, pp.3–32.

Klima, R. (1972), 'Theoretical pluralism, methodological dissension, and the role of the sociologist: the West German case', *Social Science Information*, 11, 3/4, pp.69–108.

Kuhn, T. S. (1962), *The Structure of Scientific Revolutions*, University of Chicago Press.

Lakatos, I. (1970), 'Falsification and the methodology of scientific research programmes' in Lakatos, I. and Musgrave, Alan E. (eds), *Criticism and the Growth of Knowledge*, Cambridge University Press, pp.91–195.

Lakatos, I. (1971), 'History of science and its rational reconstructions' in Cohen, R. and Buck, R. (eds), *Boston Studies in the Philosophy of Science*, VIII, Dordrecht: Reidel, pp.91–136.

McGuire, William J. (1968a), 'Theory of the structure of human thought' in Abelson, R. P. *et. al.*, op. cit., pp.140–62.

McGuire, William J. (1968b), 'Selective exposure: a summing up' in Abelson, R. P. *et al.*, op. cit., pp.797–800.

Martins, H. (1972), 'The Kuhnian "revolution" and its implications for sociology' in Hanson, A. H., Nossiter, T. and Rokkan, Stein (eds), *Imagination and Precision in Political Analysis: Essays in Honour of Peter Nettl*, London: Faber and Faber, pp.13–58.

Merton, Robert K. (1957), 'Science and democratic social structure' in Merton, Robert K., *Social Theory and Social Structure*, New York: Free Press, pp.550–61.

Miller, George A., Galanter, Eugène and Pribram, Karl H. (1960), *Plans and the Structure of Behaviour*, London: Holt, Rinehart and Winston.

Mulkay, Michael (1969), 'Some aspects of cultural growth in the natural sciences', *Social Research*, 36, pp.22–52.

Münch, Richard (1972), *Mentales System und Verhalten: Grundlagen einer allgemeinen Verhaltenstheorie*, Tübingen: J. C. B. Mohr (Paul Siebeck).

Musgrave, Alan E. (1971), 'Kuhn's second thoughts' (review article), *British Journal of Philosophy of Science*, 22, pp.287–97.

Popper, K. R. (1966), *The Open Society and Its Enemies*, London: Routledge & Kegan Paul.

Popper, K. R. (1970), 'Normal science and its dangers' in Lakatos, I. and Musgrave, Alan E. (eds), *Criticism and the Growth of Knowledge*, pp.51–8.

Reif, F. (1965), 'The competitive world of the pure scientist' in Kaplan, N. (ed.), *Science and Society*, Chicago: Rand McNally, pp.133–45.

Scherhorn, Gerhard (1969), 'Der Wettbewerb in der Erfahrungswissenschaft: Ein Beitrag zur Allgemeinen Theorie des Marktes', *Hamburger Jahrbuch für Wirtschafts–und Gesellschaftspolitik*, 14, pp.63–86.

Sears, David O. (1968), 'The paradox of *de facto* selective exposure without preferences for supportive information' in Abelson, R. P. *et al.*, op. cit., pp.777–87.

Storer, N. W. (1966), *The Social System of Science*, New York: Holt, Rinehart and Winston.

Whitley, Richard D. (1972), 'Black boxism and the sociology of science: a discussion of the major developments in the field' in Halmos, P. (ed.), *The Sociology of Science*, Sociological Review Monograph No. 18, Keele University, pp.61–92.

6

Mono- and poly-paradigmatic developments in natural and social sciences*

Cornelis J. Lammers

Kuhn substantiates his thesis (1970a, 1963) that science develops not in an evolutionary but in a revolutionary way, primarily by examples from the history of the natural sciences. What about the social sciences? Most, if not all, social scientists will readily agree about the lack of agreement among themselves about the basic tenets of their disciplines. The testimony of sympathetic and well-informed outsiders (e.g. Kuhn, 1970a, p.viii; Nagel, 1961, pp.448–9) and research evidence on differences in professional consensus between social and natural scientists (Storer, 1967; Zuckerman and Merton, 1971; Lodahl and Gordon, 1972) bear out the conclusion that the lamentations of social scientists about their disunity are not only a form of collective deference—proper to members of a young, upstart professional community when they interact with exponents of well-seasoned (and thus more respectable) disciplines—but also refer to an actual state of affairs.

This state of affairs could, of course, mean that most social sciences are still in what Kuhn called in his first essay the 'preparadigmatic stage' (Kuhn, 1970a, pp.12ff). It is also quite possible that

* This is a revised version of a paper read at the London Conference of the Research Committee on the Sociology of Science of the International Sociological Association, September 1972. My thanks for their helpful criticisms to many participants of this conference, especially David Edge, Rolf Klima, Hilary Rose, Ronald Stansfield and Richard Whitley. Parts of this essay have appeared elsewhere (in Lammers, 1972). I want to express my gratitude to R. F. Beerling, Peter A. Clark, A. W. Coats, P. van Daalen, J. K. M. Gevers, A. R. J. M. Hoefnagel, A. P. R. van Veen and C. E. Vervoort for their helpful comments on an earlier version of this chapter. My thanks also to Mrs I. Seeger for reading the English manuscripts.

such disciplines as psychology and sociology are by their very nature now and for evermore 'multiple paradigm sciences' (Masterman, 1970, p.74), 'polyparadigmatic fields' (Martins, 1972, pp.23–4) or 'proto-sciences' (Kuhn, 1970b, pp.244–5). Still, regardless of one's assumptions concerning the chances that any social science will eventually attain the status of a mono-paradigmatic concern, it may legitimately be asked: what were and are the determinants of the differential chances for mono- and poly-paradigmatic developments to occur in natural and social sciences?

In this chapter I will not approach this question from the customary philosophy-of-science point of view, however fruitful such an approach might be. The poly-paradigmatic character of the social sciences is probably not only a function of the vicissitudes of their study objects and of the deficiencies (or peculiarities) of their methods. It stands to reason that the institutional setting of the social sciences also has something to do with their plurality in paradigmatic assumptions.

In the general literature on the sociology of science most authors focus on the general cultural and/or structural characteristics of science as a social system or institution (see e.g. Merton, 1957, chs XV and XVI; Storer, 1966; Luhmann, 1968; Mulkay, 1969; Ben-David, 1971), limit their analysis to natural science (Hagstrom, 1965; Polanyi, 1962, 1967), or when they pay attention to social as distinct from natural science do so in order to stress the similarities rather than the differences between the two (e.g. Barber, 1962).

Perceptive references to the differences in institutional context between natural and social sciences are to be found occasionally in these general works (e.g. Storer, 1966, p.95; Luhmann, 1968, p.245) and quite often in essays concerning the specific predicaments of sociology and sociologists (see e.g. Elias, 1956; Klima, 1969, 1972; Rex, 1970). Nevertheless, it appears worthwhile to undertake a systematic comparison of natural and social science settings with an eye to explaining the developmental differences between the two.

Of course, given the limitations of an article, an exhaustive analysis is out of the question, so that the discussion will be of a hypothetical nature and restricted to some features of the institutional differences between social and natural sciences in present-day Western society. Furthermore, the comparison will focus on 'pure' and empirical sciences and not on formal sciences (e.g. mathematics, logics) or applied sciences (e.g. medical, engineering, administrative). Formulated positively: the subject matter of this chapter has primarily to do with the institutional forms within whose framework the main streams of theory and research development occur in such social sciences as economics, political science, psychology, social or cultural anthropology, and sociology, and such natural sciences as astronomy, biology, chemistry, geology and physics.

Before the main topic of this discourse is tackled it will be necessary to pay some attention to the concept of paradigm (Kuhn, 1970a).

Scientific conceptions and lay images of reality

Following Masterman (1970, pp.69–79), who argues that basically 'paradigm' in Kuhn's analysis stands for a 'way of seeing reality', a 'picture', a 'model' that precedes a scientific theory, I will define a paradigm or conception of reality as: a set of beliefs constructed and consciously applied by scientists with respect to aspects of reality, to the nature of and the relations between the units making up those aspects of reality, and to the methods with the aid of which these aspects of reality can be investigated.

It need hardly be said that ideas of reality are fostered not only by scientists but also by laymen. For analytical purposes, I will distinguish between scientific *conceptions* and lay *images* of reality, granting from the outset that in practice these two sets of beliefs shade off into each other. Let me define an 'image' of reality as: a set of beliefs shared by members of a non-scientific social group about aspects of reality and about the nature of and the relations between the units making up those aspects of reality.

Thus, by definition, conceptions and images of reality differ in three respects. First, conceptions of reality are to be found among scientists, images among non-scientists. Second, scientific conceptions of reality are, but lay images are not necessarily, 'constructed' and 'consciously applied'. Third, the scientific conception does and the non-scientific image does not necessarily include methods of investigating the reality concerned. In other words, reality conceptions can be viewed as a special kind of reality images, since conceptions of reality possess all the earmarks of reality images but are characterized by some additional features. The more or less deliberate 'construction' and 'application' of ideas about any kind of reality with the aid of systematic 'methods of investigation' distinguish the scientific from the lay approach, and therefore constitute the main difference between views of reality among scientific and non-scientific groups.

Lay images often form the source of inspiration for the origin of a scientific conception of reality. During the next stage of elaboration of a scientific conception, lay images springing from the conditions prevailing at that time in and around the institutions in which the scientists in question work and live, can affect the degree to which and the direction in which such a conception is worked out.

Finally, when a paradigm has come of age, and particularly when it has become generally accepted, there is a chance that it will supersede the lay image concerning a particular aspect of reality. This is the stage of diffusion of the paradigm.

The interaction between images and conceptions of reality occurs not only in the form of communication between laymen and scientists but also in the form of a *monologue intérieure* within the scientist or among scientists, for the simple reason that every scientist is at the same time a layman. Since they continuously perform other roles besides their roles as scientists, scholars and professionals, the very persons who create, develop, apply and disseminate scientific conceptions are bound to 'import' lay notions and 'export' professional notions, consciously or unconsciously, all the time.

Scientific progress

According to Kuhn (1970a, ch. 13), a science makes headway to the extent that its practitioners embrace one and only one paradigm in terms of which they devote themselves to 'normal science'. Such development is made possible by a certain degree of insulation of the scientific milieu from the cultural milieu in which these practitioners live their extra-professional lives (Kuhn, 1970a, p.164; Kuhn, 1968, pp.80–1).

In other words, Kuhn suggests that scientific progress (in the sense of mono-paradigmatic development) is due to a relative autonomy of science as a societal institution. But the degree of institutional autonomy of a science can refer to two different things: (a) the degree to which the subculture of that science regulates the working life of science practitioners more effectively than the culture(s) of the surrounding community, and (b) the degree to which the organizations 'carrying' the activities of these scientists can pursue policies guided more by their own institutional goal (developing scientific knowledge) than by the goals of external—e.g. governmental, religious, industrial—institutions.

Obviously, these two kinds of autonomy will reinforce each other and tend to increase the chances for scientific development. The first—or cultural—kind of autonomy means that the norms of science (universalism, communalism, etc.; see Merton, 1957, ch. 16; Barber, 1962, ch. 4; Storer, 1966, pp.76–86) will be rather closely lived up to by scientists so that the social system of science will function rather effectively in terms of its goals. The second—or structural—kind of autonomy implies that positive and negative sanctions at the disposal of the 'carrier organizations' will be utilized to enforce the norms of science and to attain its goal.

It appears likely, however, that relative autonomy of scientific institutions can be not only the cause but also the effect of the level of scientific progress. The more impressive the publicly visible achievements of a science, the higher its institutional prestige and power and the lower the risk for the scientific community that outside agencies can, will, or dare interfere with that community's internal affairs. Furthermore, there is less chance that highly developed scientific conceptions will be directly influenced by lay images for the very banal but nonetheless highly significant reason that laymen usually have not the faintest idea, let alone collectively sustained images, about those aspects of reality that are dealt with in sophisticated scientific conceptions. Finally, as Mulkay (1969) has pointed out, over and above the norms of science a paradigm itself might constitute a source of normative control. The more a paradigm is generally accepted and made operative in a science, the more it will serve as an inspiring 'creed' maintaining or increasing the *esprit de corps* of the scientific community concerned. Such *esprit* can in turn strengthen faith in both the norms of science and its paradigm(s)—in this way 'boosting' the cultural autonomy—and also fortify the spirit of resistance in the profession against outside interferences—in that way 'boosting' the structural autonomy.

The level of development of a science, of course, depends not only on its institutional autonomy but also on the quantity and quality of the material and human resources at the disposal of a science's 'carrier organizations'. Presumably, this material and human 'input' into the system of science affects the degree of scientific progress directly as well as indirectly via the degree of institutional autonomy. The more resources scientific organizations obtain, the greater their potential for applying positive and negative sanctions to further internal or external goals. These resources can in turn be seen as a function of the interests various societal groups have in the development of the science concerned. 'Interest' is used here in both senses of the term. It is not only the material, power, or status interests at stake for certain agencies, classes, or categories in the populace, but also the intellectual interests derived from basic values adhered to in these societal groups, that determine how much money and manpower will be available for the pursuit of particular kinds of scientific knowledge.

The interplay of the factors just mentioned should not be represented as a unidirectional chain of processes but rather as an open system of interdependent parts. We have already seen that the achievements of a science can enhance its institutional freedom. In addition, scientific output undoubtedly has feedback effects on the availability of means and on the interests involved in a particular science; and there is also the feedback effect of a science's output on its own general level of development, which is one of the defining characteristics of 'mature science'.

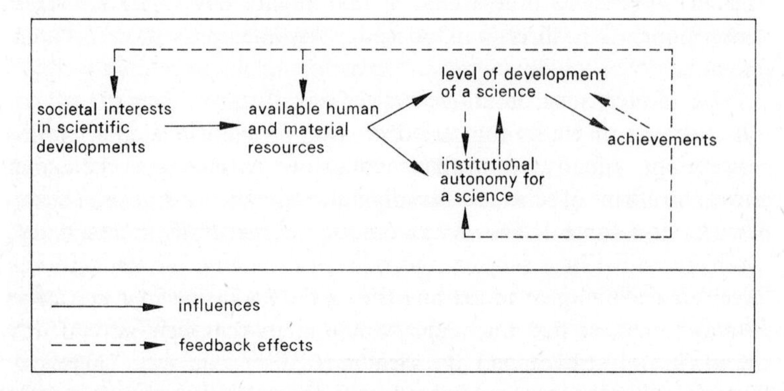

Figure 6.1 *Determinants of the developmental level of a science*

Fig. 6.1 gives a summary, highly simplified of course, of the interplay between various factors in the 'production' of paradigms. This schema enables us to compare the chances for development of generally accepted paradigms in the institutional settings of natural and social sciences.

The natural-science setting

Again applying the division of three stages in the 'life cycle' of a paradigm, let us look first at the origin. In the case of the 'pure' empirical sciences, as indicated above, there is not much chance in this day and age for lay images to exert any direct influence on the formation of new paradigms or on the creation of new theories arising within the general scope of existing paradigms, for the simple reason that these natural sciences deal with problems about which lay images usually do not exist.

Perhaps in some branches of biology the emergence of particular conceptions might be subject to anthropomorphic notions. And it is also conceivable that during scientific 'revolutions' or 'pre-paradigmatic stages' in natural science, whenever alternative conceptions appear to touch on ideologically controversial issues—think, for example, of Darwin's evolution paradigm—divergent lay images may reinforce the corresponding scientific conceptions and promote or hinder the adoption of one or another alternative paradigm (for examples, see Frank, 1951).

On the whole, nevertheless, the conclusion is probably justified that in contemporary natural science lay images rooted in societal interests and values make an imprint on new paradigms only in exceptional cases.

One wonders whether—apart from influences exerted by lay images—certain practical concerns and problems of man in his physical or social environment could also be conducive to the rise of new paradigms. Deliberate efforts by outside agencies to urge scientists to undertake the formation of new conceptions of reality are probably very rare if they ever occur at all. Usually, laymen and even professionally trained practitioners in a field will presume that basic science—that is, the prevailing paradigm in that field—could and should supply the answers to their particular problems. Therefore, outside agencies are more likely to press for the elaboration (in certain directions relevant to their pursuits) than for the origination of particular paradigms. In any case, even if outsiders did try to exert such influences, the relative autonomy enjoyed by pure-research establishments in natural science would in general prove to be a difficult barrier to overcome.

To be sure, in some cases—as, for example, Ben-David (1960) has shown for bacteriology—the initiation of a new paradigm may be the result of the practical as well as the theoretical concerns of a scientific innovator. In all likelihood, however, such subtle 'semi-outside' influences of a direct nature on the emergence of new paradigms have, at least during recent decades, contributed little to 'revolutionary' developments in natural science. As will be pointed out later on, this by no means implies that in the coming decades there will be little chance for such revolutionizing 'interferences' by the outside world.

Chances for relatively unhampered elaboration of scientific conceptions are mostly present in the empirical natural sciences, again because of the relative absence of 'competing' lay images. This is particularly evident in the socialization of the new generation of science producers (see e.g. Kuhn, 1970a, pp.165–7). The scientist-to-be is either a *tabula rasa* as regards the topics he is taught in the university or he has experienced some anticipatory professional socialization at home, at school, and in the community. Fragments of natural-science paradigms and theories have already become widely diffused and form, as a kind of 'scientific lay images', a fertile soil on which his proper professional socialization can grow. The relative ease with which a scientific subculture can be transmitted means that there is intergenerational continuity in the elaboration of professional conceptions. In other words, there is a certain institutional guarantee that successive generations of scientists can 'stand on the shoulders' of their predecessors.

Furthermore, there are few if any direct external demands for the services of the producers of pure science. Their work is too far removed from the practical problems of everyday life and what is even more important, their splendid isolation is protected by a 'buffer zone' of institutions specialized in applied science and scientific applications (for a detailed analysis of the institutional differences between pure and applied science, see Barber, 1962, pp.132–42). Schools of medicine and engineering, research and development laboratories, and the like take care of the specific and immediate demands made by society on the services of science and scientists. This division of labour between organizational units for the training of 'pure' scientists and for the creation of scientific knowledge, on the one hand, and units catering to the public demand for scientific applications and for scientifically trained personnel, on the other, entails a segregation of the two 'relevant audiences' of experts and laymen who often make contradictory appeals and observe divergent standards of excellence with respect to scientific work (Klima, 1972, pp.81–92).

The interests of various societal agencies and groups in the output of natural science (see Fig. 6.1) are heightened by the very success of scientific achievements, and this implies a tremendous pressure on the institutions generating such wonderful output 'to deliver (even more of) the goods'. Such pressures could seriously interfere with the further elaboration of the reigning paradigm and lessen the chance of broaching new paradigms. However, by assigning the task of satisfying the customer to one kind of institution and the task of developing science to another, the necessary institutional autonomy can be provided for the latter set of pure-science institutions without endangering the input of material and human resources for them.

To be sure, this state of affairs implies only that there is no direct pressure for short-term services of a specific kind from pure science. The links between institutions of applied and pure science imply that there will often be considerable indirect pressure from external agencies on long-term development in a science. Furthermore, especially since the last war governmental, industrial, and other agencies in Western countries have distributed huge amounts of money among 'pure' scientists for research in certain areas. Although in such cases no direct returns in terms of immediate interests of the financing agency or of its sponsors are expected, funding policies obviously encourage the development of some and discourage the elaboration of other paradigms. Moreover, although lay images cannot determine the content of most scientific conceptions in such fields as astronomy, chemistry, etc., lay values—e.g. about what constitutes the 'good life' or the 'good of the nation'—in all likelihood also affect the general directions in which scientists choose to develop their science.

This leads us to a final general feature of the impact of society on the development and refinement of professional conceptions of natural reality. The long-term, indirect demands concerning the general nature of scientific efforts have been and still are, on the whole, rather convergent. The control and exploitation of natural resources to increase the welfare of the population and to further national power and prestige form what Habermas (1968, pp.146–8) for one, calls the *Erkenntnisinteresse* ('the knowledge-leading interest') at the base of the development of natural science.

Finally, the diffusion stage. Here again, the paucity of lay images implies that popular notions do not often form a serious obstacle for the dissemination of fragments of natural-scientific knowledge. On the contrary, the general public and the more specific customers who seek scientific services, have been taught, in school and later on, at least some of the main assumptions of the natural-science world view, so that new ideas arising from natural-science quarters, when spread in popularized form, are easily accepted.

In this context it should be kept in mind that not only schools but also various institutions for adult education, educational programmes on radio and TV, periodicals like *Science*, amateur clubs for the promotion of the study of science, all disseminate notions derived from or built upon natural-science paradigms among significantly large parts of the populace. In other words, these institutions have a 'missionary' function, converting unbelievers and above all consolidating the faith of the converted. Diffusion of bits of natural-science knowledge via these missionary institutions therefore contributes (see Fig. 6.1) to the respect for and status of the scientific establishments and this in turn is an important factor for the maintenance of some degree of functional autonomy for these institutions (see on this point also Pinner, 1962, pp.943–4). The receptivity for natural-science achievements also contributes to the interest in and the availability of personnel and material resources for the development of these fields.

The receptivity of the public to the 'propaganda' transmitted by the missionary institutions naturally has something to do with the interests at stake for various individuals, agencies, and categories (see e.g. Handlin, 1972, pp.259–60). To a great extent there is a harmony of interests between the 'producers' and the 'consumers' of natural science. The public at large senses that somehow natural-science achievements form the basis for a rising standard of living, for national glory, and for the proliferation of fascinating gadgets to make life more comfortable and/or exciting (e.g. speed-boats, superjets, colour TV, space acrobatics). Consequently, the public at large feels—and those involved in the production or sale of the standard of living, the glory, and the gadgets *know*—that the business of science is their business. In this way the interests of indirect and direct consumers of the material achievements made possible by natural science coincide with the interests of the scientists who want money, manpower and institutional freedom to develop their conceptions.

However, there is another side to the coin. Public reactions to particular kinds of aspects of natural science often also carry overtones of fear, distrust, and sometimes even outright hostility. As pointed out by Merton (1957, pp.543–6), not only the scientific input for destructive technological products but also the general lack of concern of 'pure' scientists for the practical uses and social repercussions of their achievements, contribute to public feelings of anxiety. Moreover, the threat of scientific conceptions to traditional ways of thinking and doing (see e.g. Merton, 1957, pp.547–8; Barnes, 1972, pp.261–4) also evokes negative reactions. Finally, instead of—or even in addition to—misgivings as to the potentially destructive or disruptive implications of science, public apprehension concerning the manipulation powers (rightly or wrongly) presumed to be in the hand of experts, 'technocrats', has grown since the late sixties (see e.g. Koch and Senghaas, 1969, for a number of essays on the technocracy question).

Yet, notwithstanding such countercurrents, during the last few decades societal circumstances have on the whole fostered rather than hampered the functioning of natural science not only as a self-maintaining, but even more so as a self-expanding system. Most of the operative forces have tended to favour the institutional conditions required for particular concepts to grow into well-elaborated paradigms, and these mono-paradigmatic developments in turn reinforced the forces that sustain the institutional conditions favourable for further developments along these lines.

The social-science setting

In the social sciences, as Elias (1956, p.234) puts it: 'Men face themselves', and therefore social scientists are often much more involved than physical scientists not only in the problems of their studies but also in the problems of the 'phenomena' they are studying, at the same time tending more frequently to cherish preconceived ideas about their subject matter.

So the origin of many a social science paradigm is found in lay images that provide what Gouldner (1970, pp.29ff) calls the 'background assumptions' shaping the professional views of social scientists. Given the variety of backgrounds from which social scientists come, and given the inconsistencies and incompatibilities between lay images adhered to in these different social background milieux, it follows that from the very outset the odds are against mono-paradigmatic developments in any social science. Divergent lay images can lead to divergent scientific conceptions on all levels of theorizing, so that most social sciences constitute what Martins (1972, pp.23–4) has called 'poly-paradigmatic fields'.

The authentic or alleged resemblance of social-science paradigms to lay images forming part of some ideology or value system and the putative or real relevance of social-scientific conceptions to various kinds of societal issues have repercussions on the emergence and/or formation of new conceptions. In the first place, scientists in the human disciplines may deliberately or unwittingly take their cues from such widely disparate images and use them as material for their models, methods, etc. Furthermore, societal or human problems and concerns—even apart from the existing specific images about their nature and about ways to cope with these problems or concerns—can be a potent source of motivation for psychologists, sociologists, etc. to undertake scientific work that has little or no significance for the further development of strategic conceptions in their fields. In the second place, outsiders have less qualms in the case of social than of natural science to push for the production of new designs by social science to facilitate a fresh approach to old or new problems. They are not overly impressed by tangible benefits for the community that could be considered as 'spin-off' from social science, and are therefore less ready to accord it institutional autonomy (Klima, 1972, pp.86–7; Pinner, 1962, pp.943–4).

Lay images and external demands for 'relevant' research and 'relevant' theoretical analysis are, moreover, operative not only in the stages of inception of new trends of thought, but even more in the stage of elaboration of professional conceptions. Considerable pressure is exerted on the social scientist by the public at large, by specific agencies wanting 'service', by his students, and last but not least even by the subjects he studies, to put their problems with high priority on his agenda of professional activities (Kuhn, 1970a, p.164; Storer, 1966, p.95; Klima, 1969, pp.86–91). Since our beloved social scientist himself also forms part of many extra-professional membership and reference groups, 'their' problems are quite often at the same time 'his' problems, so that he is exposed to a permanent temptation to divert and splinter his attention among a variety of practical and general societal concerns instead of concentrating on the 'puzzles' of 'normal' science (Kuhn, 1970a, ch.4), on the tedious job of refining, testing and remodelling professional conceptions.

In this connection one should not forget that, unlike the situation in the natural sciences, establishments for the creation and elaboration of social-scientific knowledge are not well protected by a string of 'buffer institutions' which take care of external demands and provide short-term services. To be sure, there are buffer institutions of this nature in applied fields of social science, but they are frequently bypassed. Sometimes such institutions are for certain categories of potential clients too old and 'thus' not considered 'scientific enough'—e.g. in education, administration and social work—sometimes they are too young and 'thus' seen as rather unreliable—e.g. clinics for psycho- and group-therapy. In other cases clients prefer the services of unpractical, 'pure' centres of social science activity to services of applied agencies which, due to their traditional concerns with 'hardware problems', are still seen as too much oriented to the natural-science approach—e.g. in the fields of management and medicine.

I suspect that this 'defective' division of labour between 'pure' and 'applied' social science has to do not only with the relative newness of social science 'input' in applied fields but also with the nature of the social scientific contribution. The achievements of natural science are exported in the form of ideas, prescriptions, and things. The most visible and spectacular use society makes of natural science achievements is in the form of 'things' (machinery; tools; technology) and 'hardware' prescriptions, i.e. scientific methods and techniques to construct and operate the machinery. Only a limited amount of social-science knowledge finds its way into 'things' (e.g. ergonomic insights) and 'hardware prescriptions' (e.g. survey research techniques). Society consumes the output of social sciences mainly by adopting ideas (certain concepts, theories, models, approaches, either in their original or—more frequently—in a remodelled form), or 'software prescriptions' (methods of training, supervision, therapy and consultancy).

It stands to reason that the incorporation of scientific knowledge into 'things' and into 'hardware prescriptions' requires much more clearly definable expertise and lends itself more easily to specialization and professionalization than the explanation of social-scientific ideas in more or less popularized versions and the fabrication of (social) scientific 'software prescriptions'. The differentiation between raw materials (scientific ideas) and endproducts (science-based technology) is much greater in natural than in social science, so that a division of labour between the processing of the 'raw materials' and the manufacture of the 'endproduct' can flourish much more easily in the case of natural than of social science.

This definite difference in the nature of social and natural science achievements could very well constitute one of the main reasons—perhaps the one main reason—for the relative failure of the 'engineering' model in sociology that Janowitz (1972) has ascertained. It is highly significant in this context that the 'engineering model' does and the 'enlightenment' model does not recognize the distinction between basic and applied science as crucial (Janowitz, 1970). I suspect that social science 'engineers' are much more optimistic than are 'enlighteners' about the prospects of converting scientific ideas into generally applicable prescriptions, particularly into 'hardware' prescriptions.

Consequently, when exposed to external demands, 'pure' social scientists are often inclined to let themselves be diverted, not only because they share so many lay concerns, but also because they—rather than some practising professionals—are the experts on the 'enlightening' ideas which are in demand.

Another problem posed by the presence of what are often rather firmly rooted lay images about the aspects or parts of reality covered by professional conceptions, is the resistance to professional socialization on the part of the social scientist-to-be. Entering students frequently have strong convictions based on their socially supported lay images, which are by no means in tune with the professional conceptions that members of the faculty try to teach them. Unlike the student of physical science, his fellow student in social science has to a certain extent to be *de*socialized before he can be *re*socialized. In other words, whereas the novice in physical science usually undergoes a process of attitude or belief 'installation', his counterpart in social science is often exposed to a process of profound attitude or belief change. This task proves so difficult that some social scientists (e.g. Gouldner, 1970, p.403) wonder whether they should try at all to transmit their professional subculture to the next generation, and consider or prefer encouraging students to develop their own conceptions—or should one say their own 'images'? In practice, therefore, as far as the elaboration of scientific conceptions goes, there is probably less intergenerational continuity in social than in natural science.

Summing up, one can state that the high frequency with which lay images and lay demands impinge on the system of 'pure' social science, immobilizes many practitioners of empirical social science partly or wholly for salient work on the further development of this or that paradigm. These images and demands cause not only 'real' but also 'pseudo-pluralism' (Klima, 1972, pp.70–5) and stimulate conflict over 'styles of work' (Merton, 1961, pp.29–44). These divergences contribute to poly-paradigmatic developments in social science.

During the diffusion stage, lay images play a rather decisive role in the case of the social sciences. When images and conceptions about the same aspects or parts of reality square, people will sometimes give a ready ear to social science information, but at other times they will merely feel that social science is a superfluous business ('it tells you what you knew already anyhow'). In any case, social scientific achievements, if in tune with lay images, will not always have positive feedback effects (in terms of the 'system' outlined in Fig. 6.1). When social-scientific conceptions and lay images are incompatible, efforts at disseminating social science ideas will often prove fruitless and even have 'boomerang effects' for the social sciences (for examples, see Carter, 1971). On the basis of Festinger's theory of cognitive dissonance, one can surmise that when confronted with dissonant notions from social science sources, people will disregard such 'disagreeable' information and cling more firmly to their cherished images and search for new arguments in favour of their old ideas (Festinger, 1957, ch.1, pp.44–5; ch.3, pp.131–6; ch.7). Furthermore, to reduce the threatening cognitive dissonance by social means, they will look for support among those who share their images, they will try to convert others to their views and/or will start derogating those who supply them with the unwanted information (Festinger, 1957, chs 8, 9 and 10).

All this boils down to the conclusion that social science conceptions, whenever they are at odds with lay images, will tend to (a) be ignored; (b) bolster their competing lay images; and/or (c) damage the reputation of the social sciences and decrease their credibility.

Of course, social scientific conceptions do clash with lay images quite often. For one thing, any conception about—or connected with—a social problem will almost always meet with resistance from at least part of the interested public. The very fact that some societal state of affairs is defined as a 'problem' generally implies that various parties disagree about the nature and the solution of the problem. For another thing, given the fairly strong correspondence between many conceptions and images about social reality, social-scientific conceptions often smack—and are at any rate readily perceived as smacking—of certain political or ideological notions. Therefore, the opponents of the political or ideological convictions in question tend to reject such scientific conceptions out of hand.

Social-scientific results will fall on deaf ears, particularly whenever these results go counter to lay images which, for their adherents, 'stand above' any reasonable doubts. What Alfred Schuetz (see e.g. Berger, 1963, p.117) has called 'world-taken-for-granted-views' and what Rokeach (1960, pp.39ff) calls 'primitive beliefs' are lay images that are accepted as 'beyond dispute' and are therefore more or less invulnerable to scientific penetrations.

It is important to keep in mind that lay images are more often than not tied up with salient values and interests of certain social groups. This implies that in many cases discrepancies between professional conceptions and lay images about a particular issue spring from contradictory values and divergent interests. There is no overall *Erkenntnisinteresse* making for a certain harmony of interests between professionals and laymen in developing social sciences. On the contrary, there are in principle as many 'knowledge-leading interests' soliciting support from social science as there are varying interest groups. Consequently, the interests of any particular group of social scientists in spreading their insights and data about a 'relevant'—and therefore usually controversial—topic, generally coincide with the interests of some, but collide with those of other groups.

The fact that not much social-science output goes into 'things' is highly significant in this context. Social scientists cannot, so to speak, rely for their 'public relations' on the demonstrative effect of spectacular inventions and gadgets. Since they deliver only ideas and 'software prescriptions', social scientists are much more dependent than natural scientists on public receptivity to the contents of their ideas and prescriptions (see on this point, e.g. Lippitt, 1965, p.668). While natural science ideas are publicly judged and accepted not so much on the basis of their cognitive and affective content but rather on the basis of their technological and material effects, social science ideas have to be 'good'—that is, in accordance with images and interests of the publics concerned—in and by themselves.

Finally, as to 'missionary' institutions, again the social sciences are less well provided for than the natural sciences. Only some of the older, more respectable disciplines on the borderline between social and natural science (e.g. archaeology, geography), receive support from special societies, periodicals, etc. which transmit their ideas.

Secondary schools and other educational institutions, which play such a strategic role in spreading the gospel of natural science, perform a similar service for the social sciences on a smaller scale and probably with less success. Although educational dissemination of social-science knowledge may not represent much of a missionary effort, our schools could nevertheless be responsible for whatever diffuse receptivity there is in society for social science conceptions. Elements of rational and critical thought, the 'scientific' attitude, can be quite influential in reducing the number and salience of 'world-taken-for-granted-views' and 'primitive beliefs'. It could be that the secular trend toward a rising level of education in Western society implies a general 'loosening up' of lay images and will therefore eventually bring about more, even though more varied, demands for scientific conceptions of social reality. For reasons already outlined above, such increased demand is not an unmixed blessing, since it heightens the chances of disappointment with social-science output on the part of some groups at the same time that it increases the chances that other groups' needs for social-science achievements will be satisfied.

This all adds up to the conclusion that social science institutions profit much less than natural-science institutions from positive feed-back effects. Looking again at Fig. 6.1, we see that social-scientific achievements in many cases are not well received or are not received at all. Achievements that provide the public—or at least parts of the public—with dissonant information, have no effect whatsoever or even run the risk of lowering autonomy chances, of decreasing societal interests in developing the science in question, and of endangering the input of human and material resources.

Undoubtedly, these negative 'backfiring' effects of social science achievements are to some extent offset by positive reactions among those whose values or interests agree with those of the social scientists who produce the particular conceptions. But the social sciences do not form a self-expanding system tending toward consensus about and elaboration of strategic conceptions. In a sense there is no 'one' system of social science, so that the 'normal' process of social control cannot function adequately (Klima, 1972, pp.92–6). At best there are various overlapping/competing social science systems with their own 'input' and 'output' channels.

The differences in institutional conditions under which social and natural sciences operate are briefly summarized in Table 6.1.

Discussion

In many ways the foregoing analysis is rather inadequate. Let me confess a few sins of omission and commission. No attention has been paid to the university context in which most organizations for the creation and transmission of natural and social science are placed. Although most of the social sciences have waged a long struggle for academic status (see e.g. Merton, 1961, pp.23–9 on this point for sociology), today they are to some extent protected against excessively negative reactions toward some of their achievements, because the university provides a sanctuary for them. Proposals to curb their institutional autonomy or to withdraw manpower and funds from the social sciences will be absorbed to some extent by the general defences of the university institution. Whenever 'academic freedom' or other university interests are at stake, all scholars and scientists rally to the support of their threatened brethren. Thus, in a way, the expanding system of the natural sciences prevents the social-science 'system' from becoming a contracting one.

Table 6.1 *Institutional setting of natural and social science*

	empirical natural science	empirical social science
in the stage of *origin* of scientific conceptions:		
lay images are usually	absent	present
institutional autonomy is rather	high	low
in the stage of *elaboration* of scientific conceptions:		
lay images are usually	absent	present
direct, external demands occur	seldom	often
'buffer' institutions are bypassed	seldom	often
directions of external demands are mostly	convergent	divergent
in the stage of *diffusion* and application of scientific conceptions:		
lay images are usually	absent	present
'missionary' institutions are usually	present	absent
interests of 'consumers' and 'producers' of 'pure' science are usually	in harmony	in conflict

Naturally, the university context has other repercussions as well. It is often asserted that due to the institutional coexistence of social with natural sciences social scientists are liable to adopt some—if not all—of the thoughtways of their colleagues in the more prestigious 'hardware' sciences (e.g. Elias, 1956, pp.238ff; Rex, 1970; Gouldner, 1970, pp.491–2). Influence processes of this type undoubtedly occur. But one should not forget that another tradition besides that of natural science, to wit the tradition of the humanities, has always been quite influential as well with respect to the orientations of social scientists. This is not only true for European social science: in the early sixties 58 per cent of a sample of American sociologists agreed with the statement 'Sociology should be as much allied with the humanities as with the sciences' (Lipset and Ladd, 1972, p.73). It would probably be more accurate to say that disunity in the social sciences is fostered by the university context, because this setting provides social scientists with two mutually rather exclusive 'models' they can imitate. They are exposed to what Snow (1963) has called the 'two cultures', to some extent splitting up accordingly rather than forming a 'third culture' as Sir Charles kindly (1963, pp.66–7) but wrongly assumes.

Another significant omission in this treatment has been the relative neglect of 'second-order effects' of dissensus in the social sciences. By themselves, conflicts in a profession between advocates of divergent conceptions are violent enough. But once such conflicts coincide with certain diverging societal interests and values, they become even more intense and therefore form an even more serious impediment to concerted activity to develop one's science. From a sociological point of view, of course, there is no difference between laymen and professionals as regards their collective reactions toward dissonant information. Therefore, whenever they are confronted with dissonant conceptions emanating from another school, social scientists adhering to one school of thought will react as predicted by Festinger's theory. Either they will tend to disregard these other conceptions or they will double their efforts to convert colleagues, students and the public to their own professional views, and they certainly will not refrain from fiercely derogating their 'colleagues' for their 'wrong' ideas. The ensuing fratricidal disputes, to be sure, do have positive scientific significance (corrective functions, etc.) and enliven the academic scene, but they also absorb a tremendous amount of time and energy, which reduces even further the chances for each of the warring factions involved to get on with the job of elaborating their own conceptions.

Elias (1956, pp.235, 238) and Merton (1972, p.9) have observed that there is probably a connection between societal polarization and such 'balkanization' of social science. Therefore, exactly at the times when society is most in need of the service of a full-fledged and well-equipped social science, the development of such a science is most severely thwarted. The social sciences can be likened to a field hospital needed much more urgently in times of war than in times of peace, but much more likely to become incapacitated just when the demands for its services are at a peak.

Moreover, the presence of lay images and the connection between lay images and professional conceptions in the social sciences also imply 'second-order' effects in a deeper sense. The way one treats these lay images and the way one conceives of the relationship between one-self as scientist and one's subject matter, become in themselves foci for diverging 'conceptions about conceptions' and strategic bones of contention in the social sciences. Perhaps one could say that the behaviouristic school of thought in social science discards lay images even as a legitimate subject for investigation, while various 'subjectiv-istic' schools (such as the *verstehende* schools in psychology and in sociology, phenomenology, ethnomethodology) tend to attach para-mount importance to lay images, even to the point of sometimes for-bidding the study of any other aspects or parts of social reality. In a similar vein, perhaps the dispute between the neo-positivists and the critical theorists of the *Frankfurter Schule* could be viewed as focusing, among other things, on the issue of whether or not lay images should be considered either as just another kind of subject matter (the neo-positivist view) or not only as subject matter but also as *Leitmotif* for one's theoretical conceptions (the 'critical' view). In addition one might say that critical theoreticians often want the paradigms they design to be capable of immediate application as 'lay images' with emancipatory potential.

Among the 'sins of commission' detectable in the analysis of the institutional settings of social and natural science, I want to point first to the overestimation of global differences between the two kinds of science. Some social sciences are certainly more akin to natural sciences than others. Economics, for instance, has been a rather mono-paradigmatic concern (Coats, 1969), which might be due (see Table 6.1) to the fact that there has been more convergence as to the external demands on that discipline and that rather more harmony of interests prevailed between 'consumers' and 'producers' of economics than was the case with most other social sciences. In this connection it is per-haps wise to stress once again the fact that our analysis has only dealt with empirical, 'pure', and not with applied science. It stands to reason that applied natural science has much more in common with 'pure' social science than does 'pure' natural science. For example, in the medical, engineering, and agricultural sciences the elaboration and diffusion of professional conceptions is also often seriously impeded by lay images (for some examples, see Perl, 1971, and Barnes, 1972, pp.284–6).

The most serious simplification in the sketch offered above, how-ever, is its rather static description of what, of course, actually constitutes quite a dynamic development. The differences in institutional setting between natural and social science are not in a steady state, but show definite fluctuations and perhaps secular trends in the course of time. Several trends—e.g. the increases of scale—have been dealt with by Storer (1966, ch.7). In addition to such developments, which certainly have disruptive influences on the functioning of the social system of science, one could perhaps expect a growing convergence in the divergences between the two sets of sciences.

In the social sciences during the 1950s and early 1960s it looked as though a mono-paradigmatic development had finally got under way (for a documentation of this point for sociology, see e.g. Friedrichs, 1970, ch.2; Gouldner, 1970, ch.9). However, during the last five or ten years a trend toward increased dissensus about basic professional conceptions has set in (for this trend in sociology, see Friedrichs, 1970, ch.2; Gouldner, 1970, ch.10). There are no signs that this trend toward diversity will diminish in the near future.

At the same time, mono-paradigmatic developments in natural science may suffer setbacks in the future as well. If the unilinear growth during the reign of one paradigm in natural science is indeed dependent on an overriding interest—to wit, the control and exploitation of natural resources in favour of increased welfare, etc.—there are reasons to suppose that perhaps alternate conceptions in natural science will be offered to deal with the problems of environmental pollution, exhaustion of natural resources, etc. Doubts about the direction of scientific endeavours and feelings of crisis are at least clearly present among practitioners of natural science (see e.g. Zuckerman, 1971) and may give rise to divergent approaches. Likewise, the distrust and fear among the public of the unleashing of the secret powers of technology and science (the theme of Goethe's sorcerer's apprentice), which have always been present to some extent, are in all likelihood gathering strength. This could easily mean increased resistance and negative feedback effects for natural even more than for social science.

If these trends were to become dominant, there could indeed emerge in the not too distant future a certain convergence in the diverging conditions under which science as such has to operate in modern society.

Let me conclude this essay with one final, personal comment concerning the evaluation of the poly-paradigmatic nature of social science.

The comparison between the institutional settings of social and natural science provides a partial explanation of why mono-paradigmatic developments are less likely to occur in social than in natural science. Should one evaluate this as a positive or as a negative conclusion? Those who believe in the liberating power of a unified social science for the good of mankind, will probably take a dim view of the lack of prospects for professional consensus in social science. However, I do not think there is any basis for such pessimism. After all, imagine what it would be like if social scientists all agreed about the basic tenets of their disciplines and succeeded by concerted action in elaborating their paradigms in a way similar to that of their colleagues in natural science. This would certainly mean a restriction of the range of uses of social science for varying and opposing societal groups. It could even imply such an accumulation of power in the hands of one class of experts that the nightmares of *1984* and *Brave New World* would indeed come true. There is every reason to believe the truth of Lord Acton's famous statement that 'power corrupts, and absolute power corrupts absolutely'. In other words, mono-paradigmatic developments in social science could entail severe risks for any kind of democracy. That alone, I feel, is already a necessary and sufficient reason not to deplore too much the multiplicity of social scientific conceptions, because this variety guarantees that social sciences will never serve one master.

References

Barber, Bernard (1962), *Science and the Social Order*, New York: Collier Books.

Barnes, S. B. (1972), 'On the reception of scientific beliefs', *Sociology of Science*, ed. Barnes, Barry, Harmondsworth: Penguin Books.

Ben-David, J. (1960), 'Roles and innovations in medicine', *Americal Journal of Sociology*, LXV, pp.557–68.

Ben-David, J. (1971), *The Scientist's Role in Society: A Comparative Study*, Englewood Cliffs, New Jersey: Prentice Hall.

Berger, Peter L. (1963), *Invitation to Sociology: A Humanistic Perspective*, Garden City, N.Y.: Doubleday (Anchor Books).

Carter, Reginald K. (1971), 'Clients' resistance to negative findings and the latent conservative function of evaluation studies', *American Sociologist*, 6 (May), pp.118–24.

Coats, A. W. (1969), 'Is there a "structure of scientific revolutions" in economics?' *Kyklos*, XXII, pp.289–96.

Elias, Norbert, (1956), 'Problems of involvement and detachment', *British Journal of Sociology*, VII, pp.226–52.

Festinger, Leon (1957), *A Theory of Cognitive Dissonance*, Evanston, Ill.: Row, Peterson and Company.

Frank, P. (1951), 'The logical and sociological aspects of science', *Contributions to the Analysis and Synthesis of Knowledge*, Proceedings of the American Academy of Arts and Sciences, 80 (July), no. (1).

Friedrichs, Robert W. (1970), *A Sociology of Sociology*, New York: Free Press.

Gouldner, Alvin W. (1970), *The Coming Crisis of Western Sociology*, New York: Basic Books; London: Heinemann (1971).

Habermas, J. (1968), *Technik und Wissenschaft als 'Ideologie'*, Frankfurt am Main: Suhrkamp.

Hagstrom, W. O. (1965), *The Scientific Community*, New York: Basic Books.

Handlin, Oscar (1972), 'Ambivalence in the popular response to science', *Sociology of Science*, ed. Barnes, Barry, Harmondsworth: Penguin Books.

Janowitz, Morris (1970), *Political Conflict. Essays in Political Sociology*, Chicago: Quadrangle Books.

Janowitz, Morris (1972), 'Professionalization of Sociology', *American Journal of Sociology*, 78 (July), pp.105–35.

Klima, R. (1969), 'Einige Widerspruche im Rollen-Set des Soziologen', *Thesen zur Kritik der Soziologie*, ed. Schäfers, Bernhard, Frankfurt am Main: Suhrkamp.

Klima, R. (1972), 'Theoretical pluralism, methodological dissension and the role of the sociologist: the West German case', *Social Science Information*, 11, pp.69–108.

Koch, C. and Senghaas, D. (1969), *Texte zur Technokratiediskussion*, Frankfurt am Main: Europäische Verlagsanstalt.

Kuhn, T. S. (1963), 'The function of dogma in scientific research', *Scientific Change*, ed. Crombie, A. C., London: Heinemann. Reprinted in *Sociology of Science*, ed. Barnes, Barry, Harmondsworth: Penguin Books.

Kuhn, T. S. (1968), 'The history of science', *International Encyclopaedia of the Social Sciences*, 2nd ed. by Sills, David L., vol. 14, New York: Macmillan.

Kuhn, T. S. (1970a), 'The structure of scientific revolutions', *International Encyclopaedia of Unified Science*, vol. 2, No. 2, University of Chicago Press.

Kuhn, T. S. (1970b), 'Logic of discovery or psychology of research' and 'Reflections on my critics', *Criticism and the Growth of Knowledge*, eds Lakatos, I. and Musgrave, Alan E., London: Cambridge University Press.

Lammers, C. J. (1972), 'Mensen praten mee en tegen—over kritiek en krisis van de menswetenschappen', *Menswetenschappen vandaad. Twee zijden van de medaille*, eds Lammers, C. J., Mulder, M., Schroder, M., van Straten, J. and van der Vlist, R, Meppel: Boom.

Lippitt, R. (1965), 'The use of social research to improve social practices', *American Journal of Orthopsychiatry*, 35, pp.663–7.

Lipset, Seymour Martin and Ladd, Everett Carl, Jr. (1972), 'The politics of American sociologists', *American Journal of Sociology*, 78 (July), pp. 67–104.

Lodahl, Janice Beyer and Gordon, Gerald (1972), 'The structure of scientific fields and the functioning of university graduate departments', *American Sociological Review*, 37 (February), pp.57–72.

Luhmann, N. (1968), 'Selbststeuerung der Wissenschaft', *Jahrbuch für Sozialwissenschaft*, 19, pp.147–170. Reprinted in Luhmann, Niklas (1970), *Soziologische Aufklärung. Aufsätze zur Theorie sozialer Systeme*, Opladen: Westdeutscher Verlag.

Martins, H. (1972), 'The Kuhnian "revolution" and its implications for sociology', *Imagination and Precision in the Social Sciences*, eds. Nossiter, T. J., Hanson, A. H. and Rokkan, Stein, London: Faber and Faber.

Masterman, M. (1970), 'The nature of a paradigm', *Criticism and the Growth of Knowledge*, eds. Lakatos, I. and Musgrave, Alan E., London: Cambridge University Press.

Merton, Robert K. (1957), *Social Theory and Social Structure*, Glencoe, Ill.: Free Press.

Merton, Robert K. (1961), 'Social conflict over styles of sociological work', *Transactions of the Fourth World Congress of Sociology*, vol. III, pp.21–44, Louvain, Belgium: International Sociological Association.

Merton, Robert K. (1972), 'Insiders and outsiders: a chapter in the sociology of knowledge', *American Journal of Sociology*, 78 (July), pp.9–47.

Mulkay, M. (1969), 'Some aspects of cultural growth in the natural sciences', *Social Research*, 36, pp.22–52. Reprinted as ch. 7 in Barnes, Barry (ed.), *Sociology of Science*, Harmondsworth: Penguin Books.

Nagel, Ernest (1961), *The Structure of Science: Problems in the Logic of Scientific Explanation*, London: Routledge & Kegan Paul.

Perl, Martin L. (1971), 'The scientific advisory system: some observations', *Science* (September), pp.1211–15.

Pinner, Frank (1962), 'The crisis of the State Universities: analysis and remedies', *The American College: A Psychological and Social Interpretation of the Higher Learning*, ed. Sanford, Nevitt, New York/London: Wiley.

Polanyi, M. (1962), 'The republic of science', *Minerva*, 1, pp.54–73.

Polanyi, M. (1967), 'The growth of science in society', *Minerva*, 5, pp.533–45.

Rex, John A. (1970), 'The spread of the pathology of natural science to the social sciences', *The Sociology of Sociology*, (ed.), Halmos, Paul, Sociological Review Monograph No. 16 (September), pp.243–61, University of Keele.

Rokeach, Milton (1960), *The Open and Closed Mind. Investigations into the Nature of Belief Systems and Personality Systems*, New York: Basic Books.

Snow, C. P. (1963), *The Two Cultures and a Second Look: An Expanded Version of the Two Cultures and the Scientific Revolution*, New York: Mentor Books.

Storer, N. W. (1966), *The Social System of Science*, New York: Holt, Rinehart and Winston.

Storer, N. W. (1967), 'The hard sciences and the soft: some sociological observations', *Bulletin of the Medical Library Association*, 55, pp. 75–84.

Zuckerman, Harriet and Merton, Robert K. (1971), 'Patterns of evaluation in science: institutionalization, structure and functions of the referee system', *Minerva*, IX, pp.66–100.
Zuckerman, S. (1971), 'Scientific expectations: The T.L.S. lecturers', *The Times Literary Supplement*, 29 October, 5 November, 12 November.

7

'Do not adjust your mind, there is a fault in reality'; ideology in the neurobiological sciences

Hilary Rose and Steven Rose

The scientists' movement and the sociology of knowledge

Our interest in the question of ideology in the natural sciences comes from two sources, to no little extent reflecting our separate disciplines yet joint intellectual and practical concerns over these past few years. The first source is familiar enough to sociologists, coming as it does from the logical development of a renewed interest in the sociology of scientific knowledge, as against the sociology of scientific institutions or of scientists, which has dominated the field for so long. Indeed, although the Mertonian paradigm has been sustained by three decades of Columbia Ph.D.s, and it is by no means moribund, it is evident both from this meeting and other symposia (Halmos and Albrow, 1972) in the sociology of science that it is no longer the central dogma for sociologists active in the field.

The second source is perhaps less familiar to sociologists as it reflects the intellectual preoccupations of natural scientists who find themselves, as part of a crisis within science in the face of the Bomb and, more sharply, the complicity of science in the Indo-China war, to have lost faith in the automatic juxtaposition of science and human welfare (Wilkins, 1971; Rose, 1971). The traditional belief that science, because it is not ideological, must either be disinterested, or more positively, concerned with the welfare of all mankind and not limited to serve the interests of the dominant class, has been threatened. When, increasingly over the past decade, what seemed to be disinterested or even benign, such as Arthur Galston's work on plant hormones, has been press-ganged into military service as a defoliant with genocidic and teratogenic implications (Galston, 1972), then a whole range of increasingly fundamental questions have begun to be asked of the activity of science. The questions, although initially posed in moral terms of the use and abuse of science and the responsibility of scientists, have driven inexorably towards a re-examination of the epistemological basis of scientific knowledge, the increasing proletarianization and alienation of the scientists, and towards the ideological nature of science in contemporary capitalist society. It is at this point that the movement among scientists over the past few years finds itself, from a very different standpoint, at least partly sharing the concerns of sociologists with their growing interest in the sociology of scientific knowledge.

Although we have given an analytical account of this movement elsewhere (Rose and Rose, 1972), it is perhaps worth saying a little more about its development. By and large it is true to say that in Europe the movement among scientists was primarily triggered by the May Events of 1968 whereas in the United States and Britain it emerged with growing awareness of the role of science in the Indo-China war. Thus out of Europe, in particular from Italy and France, came the impulse to democratize the cultural apparatus, including the scientists' research laboratories. The élitism of science, previously accepted as an integral part of the process of doing science, became thrown into question; instead it was argued—and briefly practised, for example in the International Genetics Institute in Naples and in the C.N.R.S. laboratories in France—that decisions about research should be made by everyone who worked in laboratories, including not only the junior scientists but also the glass-washers and cleaners. Subsequent social experiments in collective research have appeared in the USA and to a lesser extent in Britain. These vary from Utopian essays into the building of a socialist science in one laboratory, through a semi-anarchist belief in the transformation of as many existing institutions as possible, to a more sophisticated analysis of the collective as a provider of constructive failures, a rehearsal for the self-management of a socialist science.

From America came a critique which went beyond exploring specific abuses of science such as the toxic nature of CS gas, the effects of napalm, the biocidic effects of defoliants or the computer-based methodology of the war (Neilands *et al.*, 1972). It began to take issue with the role of science in the service of imperialism abroad and capitalism at home. To no little extent the work of the radicals was facilitated by the hubris of the scientific profession, so that for example the creation of napalm could form the subject of a neutrally entitled book, *The Scientific Method*, by its inventor the Harvard chemist Louis Fieser (1964). Where others were to provide a different rationalization for their work, such as serving the national interest, the free world or some such, for Fieser, scientific methodology itself produced the napalm.

It was of course this elision of scientific method and social oppression, which had been postulated in theoretical terms (Habermas, 1971):

> The principles of modern science were *a priori* structured in such a way that they could serve as conceptual instruments for a universe of self-propelling, productive control; theoretical operationalism came to correspond to practical operationalism. The scientific method which led to the ever-more-effective domination of nature . . . . Today, domination perpetuates and extends itself not only through technology but as technology, and the latter provides the great legitimation of the expanding political power, which absorbs all spheres of culture.

It spoke precisely to the anxiety of the increasingly socially aware among the scientists. Some were to drop out and join the cult of anti-science seeking refuge from the 'myth of objectivity' with Roszak's shamans (Roszak, 1969). But increasing numbers were to continue to expose specific abuses of science whilst urgently working at the theoretical issues arising from their dilemma.

This brief description enables us to summarize the three preoccupations of the scientists' movement as having been:

1 The use and abuse of science, leading to ideas of personal and social responsibility in science, and the alienation of scientific work;
2 The self-management of science;
3 The non-neutrality of science, raising the question of the ideological component within the natural sciences.

It is this third issue which concerns us here.

Ideology in natural science

In the analysis of ideological components in science, the argument has begun to penetrate beneath the level of the issues of abuse or of the relatively easily criticizable problem of the gerontocratic, élitist control of science. What Joseph Ben-David dismisses as merely a chimera of the 'so-called sociology of knowledge', the movement increasingly defines as its intellectual task: that is, 'that there are regular relationships between the perspectives and motives of the social groups on the one hand and of philosophical, legal and religious or theological systems on the other. Although natural science, which is not concerned with human affairs and experiences has not been considered to be directly determined by social perspectives and motives, it may be so determined indirectly by the unstated philosophical premises of society.' (Ben-David, 1971.)

Before proceeding with our case study, we must show how we differ from Ben-David's position and clarify the nature of the ideological components of scientific knowledge. In brief, we take as given a materialist position; there exists an objectively real world of which humans are part. Our knowledge of this external objectively real world is obtained through sense data and interpreted by our brains in relationship to past experience. The very nature of the brain itself imposes a transformation upon incoming sense data; the act of comparing the transformed sense data against stored memories of other transformed sense data imposes another transformation. As a result, between the objectively real world and human knowledge thereof, there exists a series of mediations. The epistemological framework developed by the natural sciences begins as a series of procedures and rules for interpreting sense data so as to provide reliable information (that is, predictive powers) with respect to the real world. As a science develops, the rules and interpretations become assembled in particular domains, the logical coherence of which is maintained by theoretical constructs which are attempts to unify these interpretations; the past experience against which new data is being compared is selected and organized in particular ways. Thus there emerges the artificial world of human knowledge—a world of mediated sense data, a store of organized past experience and unifying theoretical constructs. What distinguishes scientific knowledge from other types of knowledge is the degree of correspondence between the artificial world and the external real world, for it is in the nature of the artificial world of scientific knowledge that it generates predictions concerning the behaviour of the external real world; these predictions can be tested in the real world and the results of such tests interpreted once again in the artificial world.

In creating the artificial world two types of inputs are involved; the actuality of the external real world and the constructs within the artificial world with which events in the real world are compared. For any scientist there are two classes of such influences in the formation of these constructs; the internal history of the subject, the domain of knowledge with its own particular organizing paradigms into which data are to be fitted, and the social history of the individual scientist himself, including his apprenticeship within the subject.

The two histories arise of course from other inputs from the external real world, which may themselves be made the subject of a different scientific inquiry; these inputs (*indirect* inputs) are distinguished from the set which directly form the subject matter of the particular science as being outside its domain and are traditionally excluded from account. These indirect inputs mould the types of prediction the scientist makes concerning the real world and the way in which the sense data arising in response to the tests of such a prediction are mediated and assimilated into the artificial world.

What forces generate the paradigm? Clearly the many past inputs from the external real world into the domain of knowledge in the artificial world, and the social histories of the individual scientists who have mapped out the particular domain. But whereas for any individual scientist the personal idiosyncrasies of his past history may affect his artificial world in a variety of disparate ways, in so far as the paradigm forms a shared part of the artificial world of many individuals, the individual idiosyncrasies become of diminishing importance, whilst the common social histories of these who share the paradigm become of increasing importance. The major indirect influence upon the paradigm in any domain of knowledge is therefore the social history of those who share it. In so far as these shared social histories represent universal features of human existence, such as fundamental aspects of human perceptual processes, brain limitations, etc., they represent a constant indirect input and a permanent aspect of the artificial world. But in so far as they represent responses to inputs from aspects of the external social world which are not universal but relate instead to particular features of relevance to a particular historical circumstance, then the inputs are themselves variable.

While science was a relatively autonomous activity within society, with little articulation into the class and state structure, the role of the variable social inputs (see, for example, Needham, 1943) was perhaps less important. This situation is no longer the case. It is a commonplace that over the last hundred years, but particularly over the last few decades, science and the state have moved closer together. Under these circumstances the major variable input into science becomes closely related to the needs of the state itself and the interests of the dominant class group within society. Thus science done under capitalism has, amongst its indirect inputs, that of bourgeois ideology. Note that in this context we understand ideology in relationship to specific class interest within society. It represents both a set of approaches to the external objectively real world and of epistemological tools, which derive from an individual's position in the dominant class within society, or alternatively, it reflects a false consciousness concerning his relationship with that dominant class. This is a more rigorous definition than the somewhat loose use of the word ideology to mean apparently no more than 'a general system of ideas' or paradigmatic framework.

When we talk of ideological inputs into science, therefore, we should be looking for ways in which the social context, internalized by the individual scientist, finds expression within his experimentation and his attempted interpretation of the external real world within his artificial world. At the same time one may note two confounding features which often cannot be distinguished from ideology in this restricted sense. First there are what might be described as the social determinants of science, the broader societal pressures towards the commissioning of research and the engagement of scientists in the exploration of the particular features of the objectively real world, that is into particular research areas. That research funding is selective and determines research directions, whilst at the same time being itself an expression of class interests within society, we have argued extensively elsewhere (Rose and Rose, 1969) and it is now virtually a commonplace.

A second confounding feature is that, so powerful a social institution has science become in the last few decades, it has in some measure come to provide its own ideology (Habermas, 1971); there has developed an ideology *of* science, which, internalized both by scientists and a substantial number of non-scientists, is itself developing profound social consequences. It is the recognition of some of these consequences which forms one of the key features of the current radicalization of scientists discussed in the first section of this paper. The ideology of science not only claims that the methodology of the natural sciences, generally understood to mean a reductionist methodology of the type discussed below, has universalistic importance, superseding all other forms of knowledge, but also denies the significance of any indirect input into the artificial world of scientific knowledge, which becomes in this conception nothing less than an increasingly accurate copy, in the artificial world, of the external real world. The ideology of science is thus positivistic. But it also has ethical overtones, claiming to provide rules for the proper conduct of human society. The only true goal for mankind becomes, in this view, the systematic extension of the artificial world of science and the matching within it, point for point, of the external objectively real world; the rationality and objectivity of science replace all else; they provide their own guide to human progress. Science, a social product, becomes both the goal and the method for all society. The clearest exponent of this viewpoint has been the molecular biologist and archetypal reductionist Jacques Monod; he has argued this case extensively in his recently published book *Le Hasard et la nécessité* (1970). In addition, the ideology of the institutions of science, hierarchical, authoritarian, sexist and, in their assumption of the superiority of Western modes of thought, racist, cannot be ignored.

In the account of ideological components within neurobiology which follows all three of these aspects, ideology expressed through the subject, ideology of science itself, and the social determinants which help shape research, are involved. In choosing neurobiology as our exemplar of indirect ideological components we are guided by a number of factors. One is the obvious one that it represents a research area for one of us, resulting in a familiarity with the research material. A second important factor is that neurobiology is, amongst the biological sciences, closest to the social sciences, thus assisting in its accessibility.

The case of neurobiology

In studying brain and behaviour, there is now available a set of increasingly powerful techniques contributed by scientific disciplines ranging from the biophysical and biochemical through the anatomical, physiological, behavioural and clinical. In addition the conceptual apparatus provided by the language of mathematics, especially that of information and control theory, has proved fertile in terms of the generation of interpretations, albeit at present not so productive of experiment. These several disciplines and approaches, no one of which is adequate in its own right, are beginning to achieve a coherent semblance of unity in the form of a new scientific field: neurobiology. It is the whole complex represented by this field that we discuss here, so as to show how each technique or disciplinary approach reveals itself, by the experiments it produces and the questions it asks of its material, to be operating within a particular framework of belief about 'the way the brain works'. This framework of belief finds expression in the terminology and metaphor adopted to describe the brain, from the hydraulic and clockwork analogies of the seventeenth and eighteenth centuries to today's computer, monkey, rat and DNA molecule analogies. Although there are three principal groups in conflict with neurobiology, which we classify respectively as *reductionists*, *vitalists* and *interactionists*, in this chapter we are concerned only with the most powerful paradigm in Western science, that of reductionism.[1]

We shall refer briefly to its principle protagonists, the type of experimental work that their approach generates and what can be said of their ideological position; in certain cases we shall refer explicitly to certain practical technologies and social consequences which flow from the work of such individuals.

The debate is really over the question of the most appropriate analytical and descriptive tools in which to explain and predict human behaviour. Reductionist tools are those which, either by simplifying the system under study (as for example by choosing a worm or sea slug, both favoured laboratory animals), or by limiting the aspects of the system chosen for examination (as for example by considering only certain forms of 'emitted' behaviour) make the experimental problem more approachable and subsumable with the general methods and theories of science developed for less complex systems, living or non-living. Reductionist tools are well established in biology, and have proved, especially over the last three decades, so powerful, that to elevate the tool into a philosophical principle has been a ready step. Indeed the apotheosis of biological reductionism is to be found among the molecular biologists such as Monod or Crick who argue essentially that in the long run all of biology is to be derived from a study of the properties of macromolecules of which the cell is composed (such as DNA) and their interactions, and may be best understood by studying the chemistry and organization of a particular bacterium present in the human intestine (*Escherichia coli*) or—even more reduced—a virus which preys upon bacteria. Indeed one of the most interesting

[1] For a discussion of the other paradigms, see Rose (1973).

developments of the last decade has been the steady invasion of neurobiology by some of the world's leading molecular biologists (e.g. Crick, Brenner, Nirenberg and even Monod).

Analysis of reductionist patterns of thoughts is made more complex by the existence in general of two principal modes of explanation in biology, which may be defined as (1) explanation of biological phenomena in terms of the chemical and physical properties of the system under study, in response to the question 'what is this system made of?' and (2) explanation of biological phenomena in an evolutionary sense, in response to the question 'how did this system arise?'. The molecular biologists' reductionism is of the form (1): that man is 'nothing but' an assemblage of molecules, etc., while another version of reductionism, favoured by ethologists and behaviourists, takes the form of (2): that man is nothing but a naked ape, larger rat or whatever. In what follows we consider each of these forms in turn.

Molecular reductionism

Molecular reductionism can be seen at its sharpest in the field of mental disorder. What is the cause of schizophrenia? Is it to be seen—as the school of 'orthomolecular psychiatry' would argue—in the absence of certain key chemicals in the brain, or in the presence of abnormal metabolites due to genetic disorders? If so, treatment is to be found by dietary modification or the development of drugs which antagonize in some way the abnormal metabolites. Following the lead of such individuals as Osmond (Osmond and Smythies, 1952) and in recent years Pauling (1968) this school argues strongly that there is an organic, brain-located *cause* for the behavioural manifestation. This belief has a long history, for at all stages at the development of biochemistry, the fashionable molecule of the moment has tended to be implicated as the cause of schizophrenia, from glutamate in the 1950s, through an abnormality of ATP metabolism in the 1960s to today's attention to the problem of galactose in the diet. Whatever the proximate biochemical cause, there is on this thesis an underlying genetic defect, a propensity to be schizophrenic. The genetic analysis of schizophrenia has been pioneered by such men as Slater at the Maudsley Institute of Psychiatry in London (Shields and Gottesman, 1971; Iversen and Rose, 1973). By contrast, sociological, social psychological, or Freudian explanations of schizophrenia stress the social and familial environment and the personal history of the individual; they point for instance to the different distribution of schizophrenia with family type and across social class. This approach is most commonly associated at the present time with the names of Laing (1965), Esteson (1970) and Cooper (1971). The dichotomy between the two approaches is complete; the paradigms meet only at the level of mutual abuse.

The ideological components within the reductionist paradigm here are apparent; the inborn view of schizophrenia at once refuses to admit criticism of social structures, such as the family and alienated work forms, whilst at the same time encouraging a manipulative view of treatment which is even more apparent when we look at the respective analyses, biochemical or social, as applied to the affective disorders such as depression. Those who argue a primarily biochemical cause of depression, such as the psychiatrist Sargent (1967, 1972), look for treatment by way of anti-depressant drugs; treatment is effective if it adjusts the depressed individual (typically a woman post-natally or around menopause) back into an acceptable social role, such as the good housewife or mother. The stability and appropriateness of the social order is taken as a natural given in this situation, and the job of the psychopharmacologist is to chemically fit people to it; it is not surprising to learn that 50 million patients were given chlorpromazine within the first decade of its use, or that 12 million barbiturate and 16 million tranquillizer prescriptions are issued a year in Britain.[1] The contrast with the politically destabilizing and non-drug-oriented paradigm of Laing, Cooper and Esteson scarcely needs stressing.

The search for a biological rationale for problems of the social order has reached new heights in recent years with a definition of a new clinical syndrome, the so-called 'minimum brain dysfunction'; a disorder experienced, so it is claimed, by pre-puberty school children. Whilst there is nothing detectably 'wrong' with the brain at the physiological level, this class of children show behaviour problems at school; they are poor learners, inattentive in class and disrespectful of authority. All these, according to Wender (1971) one of the principal protagonists of this disorder, are symptoms of a brain disorder for which the treatment is to dose the children daily with amphetamine. The response is claimed to be impressive, the children become more

[1] This is not, of course, to suggest that there is no role for drugs in the relief of human misery.

respectful of authority and quieter, less of a nuisance in class; they learn better. 250,000 school children have been diagnosed in this way on the basis of their school reports and are on daily amphetamine (or its congener Ritalin) treatment in the USA. Once again, this sort of reductionist biochemistry ignores such other social causes for inattentiveness in class as poor nutritional state, home environment—or just bad teaching.

Biochemistry is not the only brain discipline whose reductionism has both ideological and direct social significance. Physiology and anatomy have shown similar tendencies. Over recent years it has become increasingly apparent that cellular activity in particular brain regions is associated with certain behaviour patterns, so that for instance, when certain regions of the hypothalamus, a deep cellular region of the brain, are electrically stimulated by implanted electrodes in the cat or rat, then, depending on the particular cells stimulated, the animals show hunger, thirst, satiety, anger, fear, sexual arousal or pleasure. Surgical removal of these regions is associated with the reciprocal behavioural effect to that of cellular stimulation. The reductionist interpretation of these experiments is that the firing of particular cells in the hypothalamus *causes* anger, sexual arousal, etc., and, as with the biochemists, the social technologies which have emerged, notably in the hands of Delgado (1971) in the USA, have been human experimentation in which schizophrenics and low IQ patients have permanently implanted electrodes to the hypothalamus, remotely radio controlled by the doctor/experimenter, which, when turned on, are associated with sharp mood changes in the patients. Once again, the patients' anger, arousal and so on is seen as a consequence of the functioning of particular brain cells; they can be manipulated and so the patient can be manipulated, irrespective of the external circumstances which might be expected to be causes of the individual's mood. Implanted electrode studies in animals and humans can be expected to develop substantially in the next few years.

Still more revealing is the recent growth in popularity of psychosurgical techniques in the USA, and also in Britain, Japan and other countries (Breggin, 1972). The protagonists of these techniques argue that particular behavioural patterns are associated with malfunction or hyperfunction of specific brain regions, so that the appropriate medical strategy is the removal of these regions, a surgical approach which is a modification of the old prefrontal lobotomy developed by the Portuguese surgeon Egon Moniz (Hordern, 1968) in the 1930s, popular for use of schizophrenics in the early 1950s, but more recently a relatively declining treatment.

Increased knowledge of the hypothalmic centres and related regions of the limbic system of the brain (a region probably associated with fear, anger and similar emotional responses) has led to a considerable ramification of these techniques. Surgical removal of such brain regions as the amygdala has been both proposed and practised—at the rate of some 400 operations a year in the USA—to deal with individuals suffering from behaviour problems without any obvious 'organic' brain dysfunction. Such psychosurgery is intended as a pacifier, producing better adjusted individuals, easier to maintain in institutions or at home. In the USA the commonest groups of patients are claimed to be working-class blacks and women. A book by two psychosurgeons (Mark and Ervin, 1970) has recently advocated the surgical removal of the amygdala in some 5–10 per cent of all Americans, as a means of controlling ghetto violence. Once again the reductionist slogan is the reverse of that painted on the Oxford College wall; 'do not adjust your mind, there is a fault in reality.'

In these examples of molecular reductionism we see an amalgam of all those features of ideologically based science discussed above. The research paradigms do not merely dictate the experimental operations conducted, such as the search for abnormal metabolites or particular 'centres' in the brain, but have an ideological significance which lies both in determining scientific directions and in providing a powerful scientific rationale for particular social interests. But, not only do these paradigms provide ideological support for the existing social order (it is your brain that is at fault if you are disaffected); they provide a set of social technologies which help maintain that same social order.

Evolutionary reductionism

Working within a quite distinct class of reductionist paradigm, are those whose explanatory efforts are related not so much at attempting a cellular or chemical explanation of brain events but instead at interpreting the working of the brain as a system, in terms of animal models of human behaviour. This approach typifies both experimental psychologists and ethologists, albeit the schools within the disciplinary areas are themselves bitterly divided; both tend to regard molecular reductionism as giving little useful information about behaviour.

(a) *Behaviourism*

The most classically reductionist of the warring schools in this area is that of Behaviourism, dating from Watson in the 1920s. Behaviourist theory is one which is simultaneously extremely environmentalist and highly reductionist. It takes almost as a tenet of faith that all aspects of animal or human behaviour can be, and are, shaped by means of particular combinations of rewarding or aversive stimuli. At the same time however, it claims to be able to reduce all aspects of human activity to a system of 'emitted behaviours'. What is important to Behaviourism is what is measurable; events which occur within the brain and which are unobservable (intervening variables) are of little importance. The animal model for human behaviour favoured by the Behaviourist is that of a rat or pigeon in a box provided with a lever it can press for reinforcement; indeed the key Behaviourist concept is that of *reward.* This approach to human behaviour is a classic category reductionism, where all aspects of human activity, from the writing of an academic paper, through the factory production line to altruistic self-sacrifice in war or struggle, are defined as behaviours emitted in mechanistic response to past patterns of reinforcement for the individual. The Behaviourist school is sharply distinguished from other psychological paradigms, publishing its own journals and regarding as its mentor B. F. Skinner, and it is therefore of interest to examine the Behaviourist position on human behaviour as evinced by Skinner's book *Beyond Freedom and Dignity* (1972) in which he argues that all human activity is embraced within his concepts. This type of reductionism is at its worst when Skinner considers the relationship of culture to the individual, serving to control and manipulate him. He cannot see that the contradictions between individuals are themselves a part of and contained within the overall structure of society; that it is not culture as a reified abstract which controls individuals, but that culture is a product of competitive classes and groups within society. Parents and teachers manipulate and control children, as Skinner points out; but it is ignored that these parents and teachers have themselves in their turn been manipulated and controlled.

Because of this, despite Skinner's emphasis on the possibility of 'designing a culture; there is an ahistoric, static quality about his concept of society. Nowhere does he present a vision of a future culture: instead he emphasizes the 'ethical neutrality' of his techniques, applicable presumably equally to fascism, liberal democracy or socialism. Simultaneously he makes the strange error of claiming that 'no theory changes what it is a theory about.' Yet the remarkable thing about man and his society is that they are changed by theories, precisely because theories modify consciousness. In fact, because Skinner's ahistoric concept carries conviction only within the atmosphere engendered by a society of the sort Marcuse characterized as one of repressive tolerance, Skinner's position is irreconcilably conservative, and its emphasis on reward as the unifying concept for describing human behaviour is deeply ideological.

(b) *Ethological determinism*

However, the response to the Skinnerians has been the emergence of an alternative form of evolutionary reductionism, albeit generated in revulsion from the rigid techniques of the behaviourists. This is the approach of ethology, which argues that to place an animal in such unnatural surroundings as a Skinner box, or a maze made of wood or perspex, can only provide limited information concerning its capacities and behaviour—and these may be false or misleading. The alternative should be the study of animals, as far as possible unrestricted by laboratory conditions, in their natural environment and in communication with the rest of the community. Such an approach was pioneered by such people as Lorenz (1957), Thorpe and Tinbergen. The development of ethology has certainly provided a new approach to an understanding, both of patterns of behaviour and of relationships between individuals of a species, which has enriched neurobiological understanding of the complexities of social behaviour and cast new light on the functioning of the nervous system. Nonetheless, ethology—as practised by certain ethologists—has provided one of the clearest cases of ideological components amongst the neurobiological sciences.

At its most vulgar, ethological reductionism takes the form of Desmond Morris's *The Naked Ape* (1968), in which he argues that human conduct is most fruitfully interpreted, predicted and controlled in the light of studies of the ape—man's evolutionary next of kin. Whilst Morris's more extreme books, or for that matter Ardrey's *The Territorial Imperative* (1971) or the Russells' *Violence, Monkeys and Man* (1968) are by and large deplored by professional ethologists as being oriented towards the lay rather than the professional audience, they are nonetheless influential in determining further research, as for example, indicated by articles in professional journals. Whether this is a bandwagon effect—the hope of becoming a best seller, a kind of university teacher's hope for a treble on the pools—is not important here, however fascinating a speculation it may be. What is particularly apparent in these publicist accounts of ethology is the clarity with which they articulate some of the central dogmas of ethological authority. Thus the innate aggressiveness of humans is claimed directly by such distinguished experimentalists as Lorenz and Eibl-Eibesfeldt (1970) whilst Ardrey's exposition of territoriality in man derives sustenance from Wynne-Edwards's (1962) studies on territoriality in red grouse on Scottish moors, extrapolated to the human world.

Reductionist science, when a mode of operation becomes elevated by some invisible hand into a principle, is like goal displacement in organizations; a kind of explanation displacement occurs so that research which may provide an elegant account of animal behaviour is displaced into a total account of the whole human condition. Scarcely surprisingly, if man is interpreted as an ill-suppressed bundle of aggressive instincts, the formulation for social policy relates to control rather than liberation. Thus an ethologically based legitimation for conserving the social order is provided by the dominance hierarchy ('pecking order') studies; stratification is not, as most sociologists would have assumed, associated with specific societies and cultures, but reflects a genetically laid down necessity. The limitations of this particular type of ethological approach have been criticized recently by Bateson (1972) who has pointed out that not only do studies of pecking orders and dominance hierarchies relate only to particular species examined under particular conditions, but in addition, even within a group, the pecking order itself is not rigidly ordained but relates rather precisely to a particular type of experimental situation; in other situations, quite different orders may obtain, so that different dominance hierarchies may be apparent between, for example, eating activities and sexual activities.

But a reductionist ethology is one which, by definition, appropriates a set of linear and pared down analyses of particular situations, and therefore is far more prone to extract out from the richness of the experimental data, the simplistic and linear concept of a pecking order or a dominance hierarchy. In so far as the social and political beliefs of such ethologists are apparent from their writings, there are few areas of contemporary neurobiology in which ideological components stand out so sharply as in the work of ethologists such as Lorenz, Eibl-Eibesfeldt or Morris.

Genetic determinism

We come now to a class of reductionist paradigm perhaps best regarded as an amalgam of molecular and ethological reductionism (for historical reasons it is profoundly at war with Behaviourism). This paradigm we may summarize as genetic determinism; it is in general concerned with the analysis not, like ethology, of broad similarities between species-specific human behaviour and that of other organisms, but with differences between the behaviour of individual humans, or more generally between human groups. These differences are then analysed for their genetic components. The paradigm thus stresses the examination of genetic differences between individuals as expressed in phenotypic performance. Whilst genetic analysis of particular behavioural traits has been attempted with animals, for instance in rats by Tryon (1940) in the 1930s and more recently by Bovet (Bovet, Bovet-Nitti and Oliverio, 1966), and even for drosophila by Benzer (1967) it is with humans and their behaviour that most work is concerned. Although some attention has been directed to the study of the heritability of emotional traits or mental disorder, most interest has undoubtedly been associated with the work on human intelligence and its supposed class or racial distribution.

There is a long tradition of such studies, going back even beyond their generally regarded progenitor Galton, but the contemporary controversies are indissolubly linked with the names Eysenck (1971) and Jensen (1969). Perhaps what is most interesting in this resurrected debate, is the fact that relatively little at the core of the present arguments is fresh. Despite Eysenck and Jensen's claim to scientificity, their mode of analysis is similar to that which has been described as typical of the social sciences: sociologists return to central themes, whereas a natural science marches on, heedless of its past history save as a necessary precondition for its present work. A historio-sociological survey of the contribution of science to racist or class supremacist ideology would be valuable, particularly of the thirties, but certainly also during the expansion of the British Empire, together with its recrudescence during the present period when black and working-class people question the legitimacy of the stratification system.

Meanwhile, the propositions on which Eysenck and Jensen's arguments are based may be briefly listed:

1 There exists a property of humans, intelligence, which may be measured by IQ tests and which is non-uniformly distributed across the population.
2 Within British or American white populations IQ scores between individuals are more closely alike the closer the relationship (monozygotic twins closer than dizygotic twins), etc.
3 Such studies enable one to conclude that within a given population about 80 per cent of the variance in IQ score is genetically determined, about 20 per cent environmentally.
4 When examination is made the scores of working-class versus middle-class whites in Britain, or whites versus blacks in the United States, differences in scores occur which are larger than can be accounted for on the 80–20 apportioning of genetic and environmental contributions to variance.
5 Therefore, there are genetically based racial differences in IQ.

The reasons why each of these propositions is open to question, and the conclusions do not in fact follow, have been discussed elsewhere. Indeed, the issues were well dealt with during the 1930s' eruption of this debate, by, for instance, Hogben (1933) who discussed the problem of within versus between class difference and the culture laden nature of IQ scores, and Müller (1935) who demonstrated elegantly that economics, not genetics, was the key determinant. For our purposes here we may note first, the psychological reductionism (IQ measures intelligence, etc.) the validity of which had been challenged by many psychologists and is likely to be familiar ground. Second, we may note the genetic reductionism—that it is possible to parcel out behavioural attributes between a genetic and an environmental component—a supposition which is confounded by the existence of a large variety of behavioural effects which are non-genetic in character but are transmitted across two or even three generations (in rats the nutritional status of the female at birth affects brain cell number and performance even in her well nourished off-spring); these are so called transgenerational effects (Rose, 1972) but the whole procedure of the parcelling out studies can refer only to genetic potential expressed *within* a particular environment; alter the environment and the distribution itself is affected (even geneticists relatively sympathetic to Eysenck and Jensen's position, such as Thoday (1972), whose own position is reductionist enough for him to have been able to conduct an experiment in which the genetic and environmental contributions to the number of bristles on a drosophila's stomach were taken as a model for IQs, recognize this). Finally we may note (appreciated by Eysenck but not apparently by Jensen) the confusion of *within* population variance with *between* population variance; in fact within population variance says nothing about between population differences. The only way of resolving this issue would be to randomly inter-marry blacks and whites, middle-class and working-class, so as to provide a mixed gene pool, and then study IQ. Such conditions, as Bodmer (Bodmer and Cavalli-Sforza, 1970) has pointed out, are scarcely likely to obtain yet awhile.

Our concern here is with the ideological issues. If experiments which could determine the nature of the genetic contribution towards individual differences in IQ (which undoubtedly occur) are quite incapable of being performed so as to provide satisfactory evidential material, then we are forced to inquire why they are done and such emotional charge invested in them? The causes are apparent, for the ostensible results lend support for the segregated and class dominant organization of society, so as to ensure that the intelligent inherit the Earth. In any event, Eysenck and Jensen's arguments have been used in support of segregated schools in the USA and streamed education in the UK.

But the ideological output is matched by the ideological input. Friedrichs[1] has shown that Jensen's thesis is more likely to be supported by psychologists in the USA who come from the South, and less likely to be supported by those from the North or who are identifiably Jewish; it would be interesting to see the results of a similar study in Britain. Eysenck's own views on this and related topics are at least well known, and it would scarcely be necessary to dwell at such length on the point were it not that even some sociologists of science (Barnes, 1972) have quoted Jensen's as a case comparable to that of Galileo, and there have been widespread claims in the press alleging his (and Eysenck's) persecution, for their supposedly heretical commitment to the pursuit of scientific objectivity.

In fact, far from pioneering, it would appear that Eysenck and Jensen were returning to an old question. Indeed what is really interesting is the irrelevance of this debate to most current work in neurobiology, which, whether it comes from a reductionist or antireductionist

[1] 'The impact of social and demographic factors upon scientific judgment: the "Jensen Thesis" as appraised by members of the APA', to be published.

position, is not concerned with what may appear to the social scientists to be central questions. Probably the only environmentalist in a traditional sense of the word is Skinner. For most neurobiologists, individual differences are the product of the interplay of genetics and environment, and the question hinges on the inadequacy of Jensen and Eysenck's understanding of the complexity of this interaction.

Mechanistic reductionism

The evolutionary reductionists used animal models for human behaviour; a quite separate reductionist paradigm uses instead models derived from the behaviour of machines. Again there is a long history to this approach (golems were mediaeval concepts, Frankenstein's monster a nineteenth century one), hydraulic, clockwork or telephone exchange models of brain function had a long run for their money. The field of artificial intelligence proper, however, derives from the post-war development of cybernetics (Wiener, 1954) and the technology of computers. The aims are to simulate particular brain and behavioural activities using computers or related models, sometimes coupled to robot outputs, the objectives being twofold: to gain greater understanding and predictive power over brain function by providing logical models thereof and, in so doing, to generate machines with properties which will allow them to substitute for humans in a variety of ways.

Assessment of this area has not been helped by the rather dramatic pronouncements of many of its protagonists, who have described it as artificial intelligence and claimed, not only that adequate brain models were just around the corner, but that before very long computers would take over the world as a higher evolutionary form than man, reducing humans to nothing other than a temporary symbiotic role (a computer's way of making another computer) (Michie, 1970). The onward march of computers in this conception was supposed to be aided by humans in the name of either a technological imperative, or, as sometimes conceived, a sort of evolutionary imperative, to make better and higher forms of intelligence. The nature of these claims should not be allowed to obscure the relatively modest advances that have actually been made in computer modelling: a few robots which can laboriously build towers of bricks, teaching machines, chess-playing computers, and a series of largely unsuccessful attempts to produce pattern-recognizing devices and translating machines. This relative lack of success has had the useful consequence of prompting modifications of existing theories of pattern recognition and of concentrating effort into linguistic analysis. But even the enthusiasts now see it as a long time off before computers will be given the vote, a prospect once suggested by Sutherland (1968).

This modest performance may itself help focus attention on the adequacy of the paradigms of the computer modellers; essentially, they rest on the claim that human behaviour can be modelled by a set of 'logical neurons' interacting in particular networks. It would be non-materialistic to deny that, if one could build a network of logical neurons of the size and scale of the human brain (that is with 10^{10} units interconnecting by way of some 10^{14} synaptic junctions) one could generate a system with properties resembling the brain. But the logical networks of the modellers, with their few hundreds or thousands of units, each with a very small number of possible interactions, is so different in scale as to be on a level which is largely irrelevant to that of the brain; models can be made only of the brain's most straightforwardly 'logical' functions; rapid arithmetic, chess playing, or even pattern recognition—but not emotional responses, creativity, or, for that matter, poker playing. One point to be borne in mind is that the model is only as good as its programme. If the logical network of connections involved in emotional responses, creativity or poker playing were known, doubtless a model could be built which could mimic these particular outputs. Failing the theoretical framework, no model can be produced.

The value of a model lies only in its capacity to test between possible alternative theories of particular functions. Where it is not easy to predict mathematically which of two possible variants of a model will produce a given output from a given input, then to actually produce the model in hardware may resolve this difficulty. But first, a theoretical insight into the nature of the process under examination and a model of adequate complexity to deal with these processes is required.

Where this has been lacking, as in the field of artificial intelligence over the last two decades, it becomes necessary to look at the validity of the approach in terms of its sponsorship and its potential ideological role. When this is done, it cannot fail to be noted that in this field above all, there has been a preponderance of support for research from military and space agencies. This is far more strikingly true here than in any other area of neurobiology. The possible value to the military of translating machines, sophisticated computers or automated surveillance devices goes without saying; it is arguable that one aspect of the technology which has been derived from investment in this and related areas since the Second World War has been the development of the so-called automated battlefield in Vietnam, and new outputs along these lines are not hard to envisage.

The ideological outputs of the field of artificial intelligence derive primarily from its over-simplistic approach to human behaviour and performance. It is not that to make models of behaviour is itself an ideological act, but to make models of behaviour which are based on simplistic assumptions either about the behaviour to be modelled, or the nature of the model which will predict the behaviour is by definition ideological. One need not go all the way with Mumford (1971) in his distaste for 'Megatechnics' to be clear that to regard man as a machine, a soft computer, is by definition to stress his most machine-like aspects and to make him, as with machines, more biddable, programmable, subservient to the demands of the machine-maker or programmer. This must be so, until such time as the machines are really as complex and as interactive as are humans. If human intelligence is to be delimited by an analogy with the limited capacity of machine intelligence, then the prescription for the treatment of humans becomes related to the treatment of machines. They may be well serviced and maintained, but they should not get above themselves, or else, by extension, it becomes perfectly proper to give machine tools, IBMs—or even automatic lifts—the vote.

Conclusions

What we have here analysed as reductionism, that is a process where higher order phenomena are explained exclusively in lower order terms, was during the 1930s described as biologism. The issues then were mainly of race and class, with, for example, Jews very heavily under attack on the basis of a supposedly genetic inferiority; and a good deal of intellectual energy had to be spent on countering the eugenic arguments for the sterilization of the mentally, politically or racially unfit. Even when reductionist science was not challenged within itself as bad science (that is the misrepresentation, conscious or unconscious, of data) the core of the critique was the same as has been presented here, namely that 'while biology can contribute greatly to sociological study and social control, its word cannot be the last one. Sociologists and social engineers may neglect it at its peril, but equally they must master it and never let it master them!' (Needham, 1943).

We would suggest that the sociology of scientific knowledge is not merely difficult but dangerous. While the Mertonian paradigm may have supported the élitist institutions of science, the new field provides plenty of pitfalls. Lacking a substantive knowledge of genetics, neurobiology, molecular biology or whatever, the sociologist working in this field is less likely to produce the new reflexive sociology called for by Gouldner (1971) but rather a captive, reflective sociology. There are examples of this in, for example, Mullins's (1972) analysis of the phage group, for the interpretation is derived almost entirely from the phage group's own self-analysis and description, and we have little sense of sociology playing a revealing role. This is a reflective sociology which mirrors back, through the sociological journals, the self-images the scientists themselves choose to create.

For much of science the analysis of possible non-neutral components within scientific paradigms is very difficult. It may well be that it is only in a period such as the present, of social and intellectual crisis, that we can glimpse the interconnections between science and the social system. None the less, they are vital to our understanding of the nature of science. The situation in which an ideologically saturated piece of science can be described as the product of a new Galileo is however less easily dismissed. Here the contribution of the sociologist in supporting a particular piece of biological reductionism in social explanation is to provide a crucial piece of legitimation for forms of racial and class oppression which the sociologist would probably personally wish to eschew.

Other contributions to this symposium have suggested that work in the sociology of science that is limited to a sociology of scientists and scientific institutions is inadequate because it does not really provide a sociological account of science, but only a phrenological account which must, by definition, be inadequate (Whitley, 1972). The burden of this chapter is to add something else—that ignorance of the ideological implications may be a permissible plea for the natural scientists (though even this we would doubt) but that for sociologists, such ignorance is unacceptable. If the sociologist plays a role in the legitimation of oppressive ideological formulations in science, then he must do so as a conscious act.

References

Ardrey, R. (1971), *The Territorial Imperative*, New York: Dell.
Barnes, B. (1972), in 'Science and values', BBC Radio 3 Series organized by J. R. Ravetz.
Bateson, P. P. G. (1972), 'Are hierarchies necessary?' BBC Radio 3 Talk.
Ben-David, J. (1971), *The Scientist's Role in Society*, New Jersey: Prentice-Hall.
Benzer, S. (1967), *Proceedings of the National Academy of Science, U.S.A.*, 58, p.1112.
Bodmer, W. F. and Cavalli-Sforza, L. L. (1970), *Scientific American*, 223 (4).
Bovet, D., Bovet-Nitti, F. and Oliverio, A. (1966), *Life Science*, 5, p. 415.
Breggin, P. R. (1972), *U.S. Congressional Record*, H. R. vol. 118, No. 26, Washington.
Cooper, D. (1971), *Death of the Family*, New York: Pantheon.
Delgado, J. M. R. (1971), *Physical Control of the Mind: Towards a Psycho-Civilised Society*, New York: Harper & Row.
Eibl-Eibesfeldt, I. (1970), *Ethology, the Biology of Behaviour*, New York: Holt, Rinehart and Winston.
Esteson, A. with Laing, R. D. (1970), *Sanity, Madness and the Family*, Harmondsworth: Penguin.
Eysenck, H. J. (1971), *Race, Intelligence and Education*, London: Temple Smith.
Fieser, L. (1964), *The Scientific Method*, New York: Reinhold.
Galston, A. W. (1972), 'Science and social responsibility; a case history', *Annals of the New York Academy of Science*, 196, p.223.
Gouldner, Alvin W. (1971), *The Coming Crisis of Western Sociology*, London; Heinemann; New York: Basic Books (1970).
Habermas, J. (1971), *Towards a Rational Society*, London: Heinemann.
Halmos, P. and Albrow, M. (1972) (eds), *Sociology of Science*, Sociological Review Monograph No. 18, Keele University Press.
Hogben, L. (1933), *Journal of Genetics*, 27, p.379.
Hordern, A. (1968), in *Psychopharmacology*, (ed.) Joyce, C. R. B., London: Tavistock Publications Ltd.
Iversen, L. and Rose, S. P. R. (1973) (eds), *Biochemistry and Mental Disorder*, London: Biochemical Society.
Jensen, A. R. (1969), in 'Environment, heredity and intelligence', *Harvard Educational Review*, 39, p.1.
Laing, R. D. (1965), *The Divided Self*, Harmondsworth: Penguin.
Lorenz, K. Z. (1957), *King Solomon's Ring*, London: Methuen.
Mark, V. and Ervin, F. (1970), *Violence and the Brain*, New York: Harper & Row.
Michie, D. (1970), *Artificial Intelligence*, Cambridge University Press.
Monod, J. (1970), *Le Hasard et la nécessité*, Paris: Edition de Seuil.
Morris, D. (1968), *The Naked Ape*, London: Cape.
Müller, H. J. (1935), *Out of the Night*, New York: Vanguard Press.

Mullins, N. C. (1972), 'The development of a scientific speciality: the phage group and the origins of molecular biology', *Minerva*, 10, p. 51–82.
Mumford, L. (1971), *The Pentagon of Power*, London: Secker & Warburg.
Needham, J. (1943), *Time, the Refreshing River*, London: Allen & Unwin.
Neilands, J. B., Orians, A. N., Pfeiffer, E. W., Vennema, A. and Westing, A. H. (1972), *Harvest of Death*, New York: Free Press.
Osmond, H. and Smythies, J. R. (1952), *Journal of Mental Science*, 98, p.309.
Pauling, L. (1968), *Science*, 160, p.397.
Rose, H. (1971), 'Pangloss and Jeremiah in science', *Nature*, 229, p.459.
Rose, H. and Rose, S. (1969), *Science and Society*, London: Allen Lane, Penguin Press.
Rose, H. and Rose, S. (1972), 'The radicalisation of science', *Socialist Register*, (eds), Miliband, R. and Saville, J., London: Merlin.
Rose, S. (1972), in *Race, Culture and Intelligence*, (eds), Richardson, K. and Spears, D., Harmondsworth, Middlesex: Penguin.
Rose, S. (1973), *The Conscious Brain*, London: Weidenfeld & Nicolson.
Roszak, T. (1969), *The Making of a Counter-Culture*, London: Faber and Faber.
Russell, C. and Russell, W. M. S. (1968), *Violence, Monkeys and Man*, London: Macmillan.
Sargent, W. (1967), *The Unquiet Mind*, London: Heinemann.
Sargent, W. (1972), in Course Unit 17, 'Biological bases of behaviour,' Open University BBC TV.
Shields, J. and Gottesman, I. I. (1971), *Man, Mind and Heredity, Selected Papers of Eliot Slater*, Baltimore: Johns Hopkins.
Skinner, B. F. (1972), *Beyond Freedom and Dignity*, London: Cape.
Sutherland, N. S. (1968), 'Machines like men', *Science Journal* (September).
Thoday, J. M. (1972), 'Genetics and educability', *Journal of Biological Education*, 6, p.323.
Tryon, R. C. (1940), *National Society for the Study of Education*, 39.
Wender, P. H. (1971), *Minimal Brain Dysfunction in Children*, New York: Wiley Interscience.
Whitley, Richard D. (1972), in *The Sociology of Science*, Halmos, P. and Albrow, M., (eds), op. cit.
Wiener, N. (1954), *The Human Use of Human Beings*, Boston: Houston-Mifflin.
Wilkins, M. H. F. (1971), 'Possible ways to rebuild science', *The Social Impact of Modern Biology*, ed. Fuller, W., London: Routledge & Kegan Paul.
Wynne-Edwards, V. C. (1962), *Animal Dispersion in Relation to Social Behaviour*, London: Oliver & Boyd.

8

The development of sociology in the Netherlands: a network analysis of the editorial board of the *Sociologische Gids**

Wouter van Rossum

Introduction

At present sociologists of science are paying considerable attention to problems concerning the relative stages of development of different scientific fields. An important framework for these studies is provided by Kuhn's concept 'paradigm' (Kuhn, 1962, 1970). The definition of the concept of 'paradigm' is rather unclear (see for instance Masterman, 1970; Phillips, 1972); to preclude ambiguity I will use the definition Kuhn has given in reaction to his critics (using the term 'disciplinary matrix'), namely: 'symbolic generalizations, shared commitments to certain beliefs, shared values, and exemplars' (Kuhn, 1970, p.271). Lodahl and Gordon (1972), pp.57–8) have restricted the notion of 'paradigms' to 'the shared ideas of a group of scientists with regard to the problems to be investigated, the methods appropriate to their study, and the findings which are considered to be proven'. I would like to state, most clearly, that my main interest in this chapter is, consequently, in a sociological paradigm (see Masterman, 1970, p.65); therefore the term paradigm in the rest of the chapter is restricted in this sense. There remains another problem when this concept is used to describe differences in stages of development of scientific fields. This problem involves the relation between the notion of paradigm and the social structure of scientific fields. Kuhn indicates in the post-

* I wish to thank Rob Kroes and Derek L. Phillips for their comments on an earlier draft of this chapter, and R. D. Whitley for the valuable editorial comments he gave me.

script to the 1970 edition of his book, that a scientific community consists of men who share a paradigm, but he adds: 'Scientific communities can and should be isolated without prior recourse to paradigms; the latter can then be discovered by scrutinizing the behaviour of a given community's members' (Kuhn, 1970, p.176). In my opinion, Kuhn here presupposes the existence of a metaparadigm, and, especially in this quotation, is dealing only with sociological paradigms.

The study of the relation between cognitive and social development in a scientific field is interesting and important because of the obviously social nature of scientific knowledge. In this chapter, I deal with the development of a paradigm in the discipline of 'sociology', and more specifically 'sociology in the Netherlands'. A brief outline of the history of Dutch sociology will be helpful here.[1]

Sociology had come to an early stage of development in the Netherlands, especially through the influence of R. Steinmetz (1862–1940). At the University of Amsterdam he founded a 'School' of so-called 'sociography'. Sociography was intended to be a science different from sociology with the main goal of describing 'social facts'. In his conception, sociography should be the science which could supply sociology with her empirical data. For him, sociology was a fully theoretical science. This strict separation between theory and research led to a completely separate development of both 'sciences', in which sociology lacked empirical data and sociography a theoretical background.[2] Around the Second World War sociography was considered an unfruitful kind of research, merely providing descriptions of 'social' characteristics of geographical entities.

In reaction to this situation, a new generation of sociologists attempted to advance a new conception of sociology. Van Doorn (1964a, p.50) describes the situation in the early 1950s as a conflict of generations: older and younger sociologists sharing the same inheritance, but with different evaluations of it. Most certainly, however, some sociologists of the older generation also criticized sociography (see for instance den Hollander, 1948). This conflict was expressed in November 1953 in the founding of a new sociological journal, the *Sociologische Gids* (Sociological Guide), as the mouth-piece of the

[1] For a more thorough investigation of this history, see van Doorn, 1964a.

[2] There is a difference between the Dutch (Steinmetzian) conception of 'Sociography' and the German conception of this science. In particular the strict separation between theory and research has been heavily criticized by German sociographs (van Doorn, 1964a, p.39).

younger generation, taking its place beside the already existing journal *Mens en Maatschappij* (Man and Society) which was the influential communication channel of the older generation. The main goal of the founders of the *Sociologische Gids* was to extend and deepen sociology. It is not surprising, considering the history, that one of the means (among others) to reach this goal in the founders' definition was the interpenetration of theoretical-sociological thinking and social research.

I will try to trace the developments in the structure of those who edit and have edited this journal, and their scientific activities, and to elucidate the process of choosing a paradigm in a special scientific field. To do this, I will first concern myself with the social structure which the editors of the journal represented in the years between 1953 (the first issue of the journal appeared in November 1953) and 1971. To describe this structure I borrow the concept 'social network' from social anthropology. In the network, I will deal with two types of relations between the editors. Following that, a method is suggested to find within this larger structure of editors, groups of scientists who had more active relations among one another than with the other people in the large structure (the amount of activity of relations is here indicated by the incidence of co-authorships). After this excursion into the social structure of the editors, I will deal with the development of their scientific ideas in an analysis of their articles published in the journal *Sociologische Gids*.

The social structure of the editorial board

In Mitchell (1969, pp.1–50) and Bott (1971, pp.314–30) interesting reviews of definitions and conceptions of social networks are given; both indicate that this concept is used in many ways. Still, it is possible to find a basic set of ideas implicitly or explicitly held by anthropologists about the concept. A social network is, then, a set of people among whom relations can be discerned; characteristic of this set is that no exact boundary of the set can be indicated and that the relations among the persons in the network can be actual (that is factually existing) or potential. Furthermore, relations can be of a simplex or a multiplex nature (Bott, 1971, p.311). In the case of the former, the people in the network will have particular, shared interests; whereas, in the latter, the relations are more all-embracing. So Bott defines a network (as distinguished from 'groups' in the sociological sense) as a formation in which some, not all, of the component individuals have social relationships with one another, while these component units do not make up a larger social whole; they are not surrounded by a common boundary (Bott, 1971, pp.58–9).

The structural characteristics of a group of scientists editing a scientific journal, and most certainly a journal with so clearly a programme of its own as the *Sociologische Gids* in Dutch sociology, can be conceived as a network formation. The special relation of the joint editorship is in that case an expression of shared interests (in the development of sociology in the Netherlands). This, of course, is most pronounced in the years directly following the founding of the journal.

Another structural characteristic, the absence of exact boundaries, fits the editorial board structure. In the first place, during the years that the journal has existed, continual extensions and changes of the board have taken place; while otherwise the connections of the editors with other sociologists outside the editorial board, sometimes more frequent or more important than the relations inside the board structure, prevent such a board from being a real 'social whole'. Finally, the joint editorship of the journal does not necessarily imply that one editor has (or had) direct contacts with all the other editors, although actual relationships can exist for some of the editors.[1]

In Fig. 8.1, all forty-six sociologists who have been editor for some period of time are represented. Horizontally the years are marked; vertically the forty-six editors are indicated. For each editor is indicated, by a line, the time he is, or has been, on the editorial board. Only one of the editors (number 2) has been on the board from November 1953 until 1971.

Furthermore, a close look at this figure reveals that there are a few years in which the board changed radically, viz. 1958, 1959, 1965, 1967 and 1970. The year 1959 is not considered here, because the change on the board in that year was especially meant to involve Belgian sociologists in the journal.[2] In 1958 the board was reorganized and extended with new, younger members. This extension of the board was, according to an editorial comment in the first issue of 1958, accompanied by a change of the editorial policy. The change of

[1] I got the impression from personal communications with some of the editors that frequent mutual contact was more characteristic for the early years than for the later years.

[2] Some of the editors (viz. 8) have taken doctoral exams (or Ph.D.s) in a few Belgian universities (Leuven, Ghent); a consequence of the editorial policy to involve Belgian sociologists in the journal.

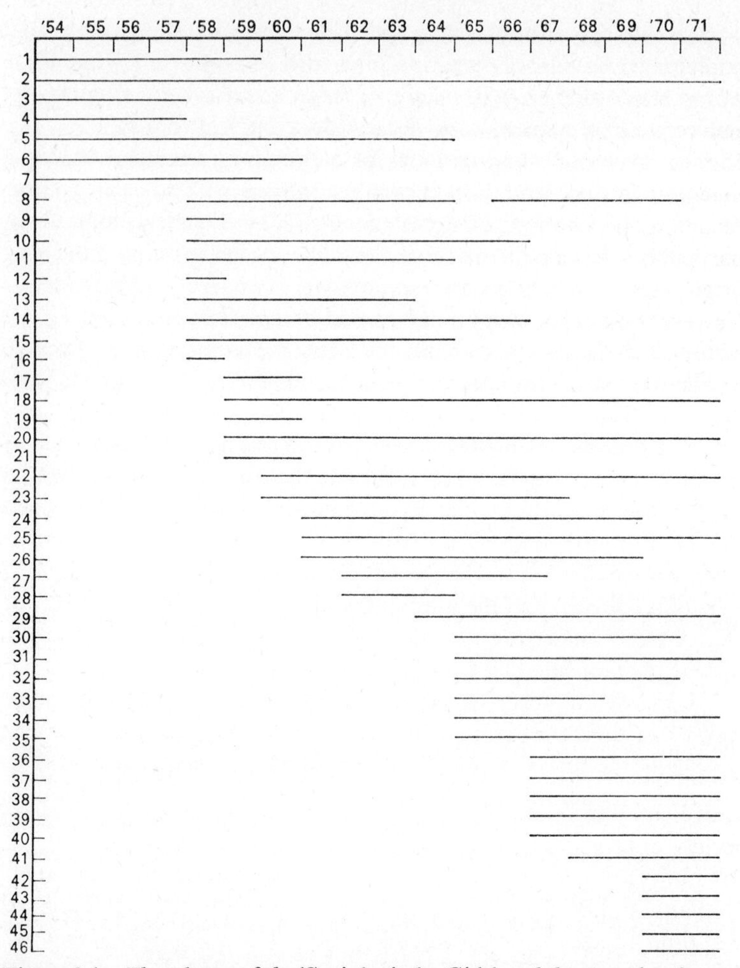

Figure 8.1 *The editors of the* 'Sociologische Gids' *and the time they have been on the board (1954–71)*

policy was not so much a radical turning point as an explicit elaboration of the programme held by the first editorial board. In this comment, it was stated that the development of sociology and social research in the Netherlands was not satisfactory; the following were mentioned: methodologically weak social research, poor organization structure, infiltration of social policy interests in the research sector, and the lack of funds for basic research. The *Sociologische Gids* had to fulfil a critical function in all these things.

The changes of the board in 1965, 1967 and 1970 were not accompanied by these kind of comments, although certainly the rejuvenation of the board will have been one of the main reasons for changing. Before we turn to examining the extent of the relations between the editors, it is interesting to look for common characteristics of the members in the board. Among others, there are I think two factors related to the sharing of interests in sociology in general and, more particularly in special fields in sociology, namely same academic origin, and same affiliation with universities as faculty member. Correspondence in one of these (or both) factors between some of the editors does increase the chances of actual collaboration in research or other scientific activities.

The academic origin of the editors and their affiliation with universities as faculty members

A basic assumption of this study is the correctness of the thesis that the development of science in general and sociology in particular is not only a question of changing ideas (or, as Popper would call it: development in the third world of truth and objective knowledge; 1968), but that these changes are accompanied by, and most certainly influenced by variations in the social structure of scientists (Crane, 1972).

Social scientists from the same academic institutions tend to have the same scientific conceptions. For instance, Heyl (1968) considers the influence of relationships among a group of scholars at Harvard University during the 1930s and early 1940s, on their conceptions of society in terms of a system with 'built-in homeostatic tendency'. In her study, she investigates the influence of L. J. Henderson on Parsons, Homans, and Brinton, who, with others, formed the 'Pareto Circle' at that time. In a study by Curtis and Petras (1970, p.209), discussing the different approaches in the field of community power, the influence of universities is also mentioned:

> It is difficult to find a scholar recently affiliated with the Yale University political science department who has not subscribed to the pluralist perspective and decisional (or later combined) approach. On the other hand it is also difficult to find a 'disciple' of C. W. Mills or Floyd Hunter, or a researcher recently affiliated with the Michigan State University sociology department who has not used some form of reputation and institutional analysis in studying community or national power.

This influence of a particular intellectual academic background is the more important for the development of a scientific discipline when the shared academic background is a source for a well-knit network of formal relations among scientists (see, for instance, Reynolds *et al.*, 1970). The same affiliation with a university is indicated by van Doorn (among others) for the school of 'sociographs' and the University of Amsterdam (van Doorn, 1964a). In the introduction, I have already pointed to the importance of Steinmetz who lectured in Amsterdam (as well as at Leyden and Utrecht) for the development of sociography at that University. It is, therefore, no surprise that the reaction to this scientific direction of a younger generation of sociologists was especially localized in the same University. As van Doorn says (1964b, p.41): 'The founders of the *Sociologische Gids* were younger people of the same age, and often friends, and for the most part Amsterdam people.' At the beginning the *Sociologische Gids* was a journal run by people who originated from the University of Amsterdam. Table 8.1

Table 8.1 *The proportion of people on the board of the* 'Sociologische Gids' (*1954–71*), *who have graduated from the University of Amsterdam*

board composition in	proportion graduated from Amsterdam	board composition in	proportion graduated from Amsterdam
1954	0·40	1965	0·33
1955–57	0·57	1966	0·37
1958	0·79	1967–68	0·48
1959	0·63	1969	0·50
1960	0·44	1970	0·58
1961	0·47	1971	0·68
1962–64	0·41		

gives the proportion of persons in the board who originated from the University of Amsterdam[1] for all the board compositions during the years that the *Sociologische Gids* has existed. In the Table, the other universities in the Netherlands are not included because, leaving out of consideration the persons who have graduated from Amsterdam, the other editors originated from all the other universities.

[1] The university in which the doctoral and not the Ph.D. exam was taken has been chosen as an indicator for academic origin; because in the Netherlands the doctoral exam is usually considered to be the end of the academic training. Consequently, the Ph.D. degree, if taken, is done later in the academic career (see Lammers and Philipsen, 1966).

Considering the whole group of forty-six editors, twenty-two (47 per cent) persons took their degree in the University of Amsterdam. This gives an indication of the importance of one particular academic origin in the recruitment of new editors. This finding is especially surprising, considering the distribution of academic origins of all sociologists in the Netherlands. According to Westerdiep (1970) 4·9 per cent of all Dutch sociologists took their degrees in the University of Amsterdam.[1] To gain insight into the changes of the board structure, concerning academic origin, a closer look at the already noted years 1958, 1965, 1967 and 1970 is given. The most striking fact in the Table is the increase of the proportion in the year 1958 (+0·22). In that year nine new editors joined the board, of whom seven persons had graduated from the University of Amsterdam. This is the more striking because in the later alterations of the board never was such a heavy emphasis put on the recruitment of new editors with Amsterdam University degrees.

There is, I think, a good reason for the clear difference between the extension in 1958 and the alterations of the board in the latter years. Scientists who are trying to find a new discipline or to change an existing one radically, will not have the opportunity to choose their partisans out of a wide circle. In such a situation similar academic origin is perhaps the major way to guarantee that the 'élan' and cohesion is maintained. This criterion will be less important when the programmatic phase is over. Another explanation for the more differentiated origins of the editors in the later years is what I would call the 'widening' of the original network. The first members of the board formed a 'clique' located in Amsterdam. In the early sixties when sociologists got their academic recognition (in 1963 separate faculties of social sciences were founded at the Dutch universities, in which sociology became a separate field of study), the original clique was scattered over more universities. In consequence of this, the number of potential relations grew considerably, which can be an explanation for the differentiation of origins of the latter editors.

In Table 8.2 the widening of the network is elucidated. In this Table is provided, for the composition of the board in several years, the

[1] Of course, this overall figure is only a rough indication. Actually, the proportion of Amsterdam sociologists in the total population of sociologists in the Netherlands is greater, when considering the graduates in the 1950s. In the 1960s the proportion decreased. So it is not surprising that the *Sociologische Gids* in origin can be conceived as a typical Amsterdam journal, but that it remained so as well.

Table 8.2 *The affiliation of editors of the* 'Sociologische Gids' *with universities (for the board compositions in 1954, 1958, 1965, 1967 and 1970)*

university*	1954	1958	1965	1967	1970
Amsterdam	1	6	4	6	6
Amsterdam (Free Univ.)	—	—	—	—	—
Leyden†	1	4	4	3	2
Rotterdam*	—	—	2	2	2
Tilburg*	—	—	—	—	1
Utrecht	—	—	—	1	2
Nijmegen	—	—	1	2	2
Groningen	—	—	1	1	3
Wageningen*	—	—	1	1	1
Belgian Univ.§	—	1	4	4	3
others‡	3	4	—	—	—

* Rotterdam, Tilburg are actually Economic Schools; Wageningen is an Agricultural University.
† In the University of Leyden is included the NIPG-TNO.
§ Viz. Leuven and Ghent.
‡ Editors working at research institutes outside the university.

affiliation of the editors with universities as faculty members.[1] Compared with the year 1953–4, when only Amsterdam, Leyden and research institutes outside the university are represented, the last year (1970) offers a picture in which all universities (except one) are represented in the editorial board of the *Sociologische Gids*. Still 59 per cent of the editors constituting the board in 1970 studied at the University of Amsterdam.

Activating the relations in the network: co-authorships of editors

In the foregoing, it is frequently stated that the structure of the editorial board of the journal could be represented as a social network structure. Some global characteristics have been treated, and from

[1] Data with regard to the affiliation with universities as faculty member, has been offered by the yearbooks of the VSWO (Vereniging van Sociaal-Wetenschappelijke Onderzoekers; The Association of Social-Scientific Researchers), for the year 1970–1 the yearbook of the NSV (The Dutch Sociological Association).

that, the conclusion is justified that, in any case, relationships existed between the editors, in the early years more tightly. Later the network expanded. Still, when we turn to consider the board structure as a real network of relations, more emphasis has to be placed on actual contacts between the editors. The latter becomes especially important when we try to operationalize through this network-concept the 'scientific community on the lowest level' (Kuhn, 1970, p.177).

Kuhn distinguishes three different notions of the concept 'scientific community'; in the most universal sense, it indicates all the scientists in a particular discipline, e.g. natural sciences, social sciences; at a lower level, it indicates the main scientific professional groups, in which subgroups can be discerned. At the lowest level, the concept scientific community can be used to indicate the scientists in a particular field who have many contacts with one another or are actually working together. 'Communities of this sort are . . . the producers and validators of scientific knowledge. Paradigms are something shared by the members of such groups' (Kuhn, 1970, p.178). In this chapter, I will use as an indicator of the active relations between the editors of the journal, their formal collaboration in research and other scientific activities, embedded in co-authorships of books and papers for the years 1954–71.[1] Of course, this indicator gives no credit to all other kinds of relations between the editors. A more thorough investigation would involve all the other types of ties and also the informal communication between the editors. Still, I think these co-authorships will give a valid representation of the extent of relationship between two editors, although probably the network discovered here is only a part of the original one, one most likely giving the most important linkages of the latter.

In Fig. 8.2 are given all co-authorships of members of the board with one another. Seventeen editors have, in the period between 1954 and 1971, worked together with other editors at least once, in publishing articles and/or books. There are several points of interest in this Figure, namely (a) the structure of this network, (b) the 'generation' differences between collaborators in the network and (c) the periods in which most collaboration occurred. A morphological characteristic of networks is the density of the network. This concept refers to the proportion of existing relations in the network as compared with the

[1] Articles published in the Dutch journals *Mens en Maatschappij* and *Sociologische Gids*. Books as far as available in the library of the Sociological Institute of the University of Amsterdam and the University Library in Amsterdam.

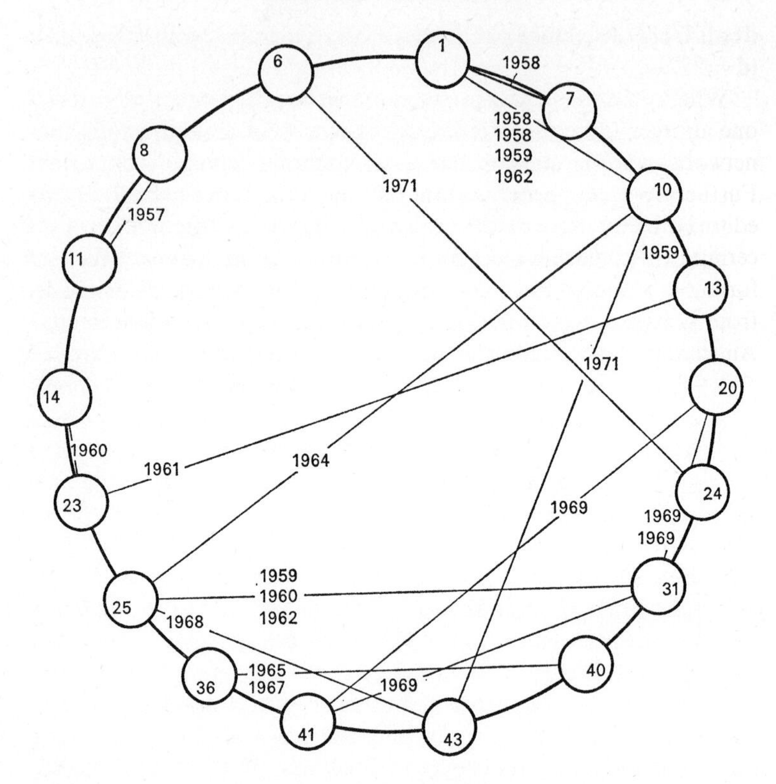

1960 year of publication of a book, co - authored by two or more editors

1960 year of publication of an article, co - authored by two or more editors

Figure 8.2 *Co-authorships of editors of the* 'Sociologische Gids' *with other editors*

total number of possible relations.[1] In the network represented in Fig. 8.2 this density is not very high (d=0·11). One of the reasons for the low density of this network is the occurrence of a few dyads in it, consisting of editors without co-authorship relations with other editors (viz. dyads 8–11, 6–24 and 36–40). When we remove these

[1] In social anthropological literature this characteristic has been named quite differently: 'connectivity', 'connectedness', 'interconnectedness', etc. In a methodological paper Barnes (1969) clarified the concept. A measure for the density of a network can be:

$$d = 2.m/n\,(n-1)$$

in which: m = actual relations among elements in the network
n = number of elements in the network.
A similar measure is used by Crane, 1970a.

dyads from the picture, the density of the remaining network increases (d=0·22).

What remains is a small, close-knit network of editors related with one another, directly or indirectly by means of co-authorships; this network can be called in terms of Mullins (1968) a 'strong unit'. Furthermore, it appears that the editors in this 'core-network' of the editorial board have several other characteristics in common. Concerning their academic origin, seven editors in the core network are former students of the University of Amsterdam (two are descended from Leyden, and two persons have degrees from the Free University Amsterdam, and Utrecht).

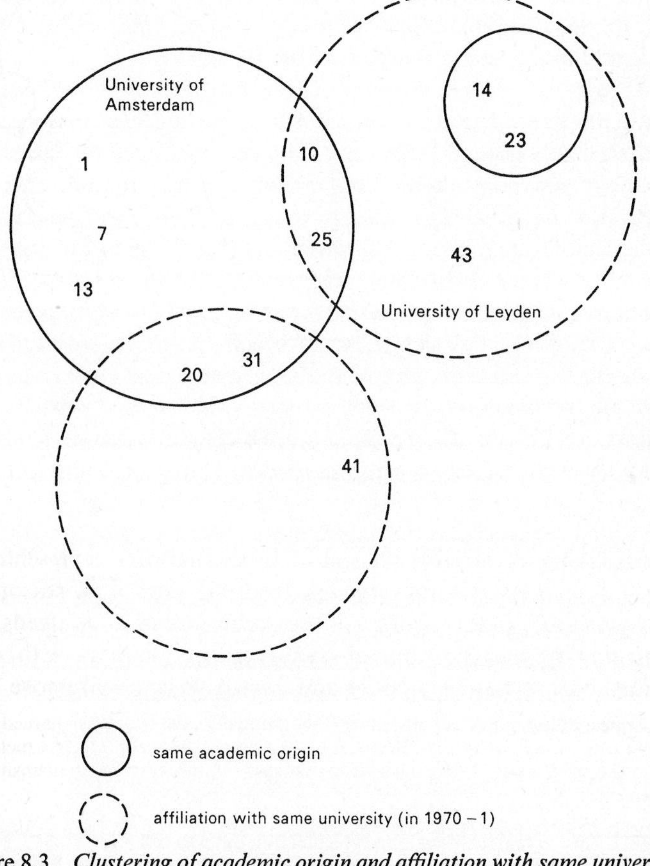

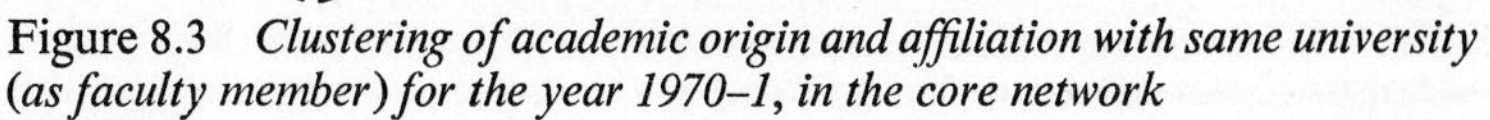

Figure 8.3 *Clustering of academic origin and affiliation with same university (as faculty member) for the year 1970–1, in the core network*

This clustering of academic origins is overlapped by a clustering of academic affiliations as faculty members at several years. As an example, in Fig. 8.3 clusters in these two dimensions have been represented (for academic affiliation the year 1970–1 is taken). Apparently, the co-authorship relation is not the only type of relation, connecting these editors. When date of entry in the editorial board is taken as some kind of indicator for belonging to a certain generation of sociologists, most of the members of the core network can be reckoned among the earlier editors of the *Sociologische Gids*; only three persons have entered the editorial board after 1963.

This datum is especially important in the light of the periods in which most collaboration occurred. It is obvious, considering Fig. 8.1, that most collaboration occurred in two periods, namely around 1958, and much later by the end of the 1960s.

We have called the early years of existence of the *Sociologische Gids* a 'preparadigm' period. For a number of reasons, this period ended approximately around 1963. I have already mentioned the fact that in that year sociology in the Netherlands got her formal, academic recognition as a separate science by the founding of sociology faculties at the Dutch universities. Of course, the founding of faculties in a particular science is not a sufficient condition to end a preparadigmal period in a science, but at the same time some textbooks on sociology and on methods and techniques of social research were published, indicating the fact that agreement about sociological knowledge grew in the Netherlands.[1]

Returning to our discussion about the collaboration, it is obvious that collaboration between the editors in writing articles and books occurred before and after this turning point. In this perspective, it is interesting to investigate the nature of the joint efforts of the editors in the core network.

Because of the fact that there can be expected a high correlation between the scientific activities of this group of editors and the development in the editorial policy of the *Sociologische Gids*, I will first deal with the latter, after which I will look at the scientific contributions of the core network.

The editorial policy of the 'Sociologische Gids'

Scientific journals play an important role in the social system of

[1] This indicates, in the words of Masterman (1970, p.65), the existence of an 'artefact or construct' paradigm.

science. A scientific journal offers to scientists the opportunity to communicate their research results to other scientists. Through the use of so-called referee systems by means of which decisions are reached with regard to publishing certain articles and refusing others, they control access to the communication channel (Hagstrom, 1965, pp.202–4; Crane, 1970b; Whitley, 1970). This procedure can be called a sort of 'unobtrusive' editorial policy. A more 'obtrusive' type of editorial policy, perhaps less usual, could be that members of the editorial board of a scientific journal ask people to send in articles or even to write articles.[1] Executing a policy for the acceptance of certain articles and the rejection of others is only possible when sufficient copy for the journal is available.

The problem for both Dutch sociological journals has always been, however, a lack of articles to publish. An investigation of the editorial policy of the *Sociologische Gids* through the study of the nature of the accepted and refused articles is, therefore, really impossible. The apparent impossibility of this research strategy made it necessary to look for other indications of this policy. One opportunity in the case of the *Sociologische Gids* is the analysis of the editorial comments appearing in each issue of the *Sociologische Gids*. Rather than analysing all articles in the journal, another strategy would be to select those written by the editors themselves in their own journal. If a journal has a distinct editorial policy, then most certainly the elaboration of this policy should be clear in the articles written by the editors.[2]

For the analysis, I combined these two alternatives. From the editorial comments, especially from the first issues, I distilled the ideal programme of action (see *Sociologische Gids*, editorial comments, 1953–4, p.1; 1958, p.1; 1960, p.1; and 1964, p.105). This part of the analysis gave me a kind of 'code-sheet' by which I could investigate the articles written by the sociologists who were, at one time or another, members of the editorial board. Through this procedure, I

[1] In a personal communication one of the founders of the *Sociologische Gids* told me that the latter was sometimes applied by the board in the first years.

[2] We can only follow this strategy, of course, when a good proportion of the articles in the *Sociologische Gids* were actually written by editors. It was found that 37 per cent of the articles in the *Sociologische Gids* (between 1954 and 1971) were written by people who were once editor of this journal. The influence of the editors with regard to the proportion of articles in the *Sociologische Gids* decreased in the most recent years, according to the following figures: 46 per cent in 1954–6, 55 per cent (1957–9), 44 per cent (1960–2), 42 per cent (1963–5), 20 per cent (1966–8), and 25 per cent (1969–71).

could find if, and to what extent, the aims set forward by the founders and other early editors were accomplished. The elaboration of the editorial programme of action is especially interesting in the perspective of the correlation between the changes in the social structure which the editors represented and the development of their sociology, here indicated by their articles published in the *Sociologische Gids*.

The main goal of the founders of the *Sociologische Gids* was very generally stated as the addition and deepening of sociology (ed. comment, November 1953, p.1). In this particular comment, they also mentioned more precisely the fields in which this programme has to be worked out; namely (a) a clear-cut apparatus of concepts; (b) exact and appropriate research techniques; (c) a reciprocal penetration of theoretical-sociological thinking and social research; and (d) over and above these fields was mentioned the relation between the science of sociology and social policy.

These four fields offered a framework to analyse the articles of the editors. During the analysis it became necessary to add a fifth category, namely (e) the clarification of the boundaries between sociology and the other (social) sciences.

These five sub-goals can be realised in a number of different ways by publishing articles. First of all, the articles can be of a normative nature, giving rules as to how it ought to be according to the opinion of the editors; I would term such articles programmatic. Another way of realising these goals can be the publishing of articles, which are considered to be good examples of the ideas about sociology held by the editors; these articles elaborating the programme I would call exemplary. Both categories were subdivided; the programmatic articles according to the five subgoals of the editors:

(1a) articles giving criteria for the relation between sociology and social policy;
(1b) articles giving criteria for the clarification of the boundaries between sociology and the other (social) sciences;
(1c) articles giving criteria for the reciprocal penetration of theory and research;
(1d) articles giving criteria for exact concepts;
(1e) articles giving criteria for the methods and techniques of social research.

In the category 'exemplary' articles a classification was made in: (a) theoretical articles (which were further subdivided in (a.1) 'Grand Theory' and (a.2) 'Middle Range Theory'); (b) theoretical-empirical articles; and (c) empirical articles (following Lammers and Philipsen, 1966).[1]

The hypothesis can be stated that in the early years more of such programmatic articles would be published, while only later in the journal would the proportion of exemplary articles increase. The publishing of such programmatic articles (accompanied by disputes about them) is, in my opinion, an empirical indication of a pre-paradigm period in a science (in this case in sociology in the Netherlands); whereas, especially, the occurrence of articles in which results of research activities are given, will indicate that apparently groups of scientists have paradigms, enabling them to solve problems in performing research. One of the research goals of this study was the discovery of a scientific community sharing a paradigm; before we can answer that question, we have to consider the dispersion over the years of the several types of articles.

In the foregoing, it was stated that the preparadigm period in Dutch sociology ended approximately in 1963. I advanced two arguments in favour of this thesis; the appearance of textbooks, and the formal, academic recognition of sociology in the Netherlands. Now, we can see if a third argument can be added, namely the nature of published articles in the journal *Sociologische Gids* (Table 8.3). For that reason in Table 8.3 the time-variable is dichotomized in the period 1954–63, and the period 1963–71. The most striking point in Table 8.3 is, of course, the obvious fact that the idea that the year 1963 has been a milestone in the development of Dutch sociology is supported by these data; the programmatic phase in the development of the *Sociologische Gids* was ended in 1963; after that year practically no programmatic articles have been written.[2]

A point of clarification is needed for the three arguments advanced explaining the turning-point in Dutch sociology around 1963. Actually, the three arguments are related. The important textbooks

[1] Methodologically, the procedure is a rather poor one. It would have been better if the classification of the articles in these categories was done by a group of 'judges', people originated from the sphere of professional sociologists. For reasons of time and money, I could not follow this plan. Maybe it is still useful to explicate especially the distinction made between theoretical-empirical and empirical articles. The former are articles in which theory, explicitly a theory, or part of a theory, is being tested. The latter refers to articles in which empirical descriptions are given of parts of social reality, without reference to special social theories.

[2] Actually, there is not such a radical turning-point in the nature of the published articles as suggested by the dichotomization of the time variable. In fact, there was a gradually diminishing proportion of this kind of programmatic article after 1960.

Table 8.3 *The nature of articles published by the editors of the 'Sociologische Gids' in their own journal in the period 1954–63 and the period 1963–72*

	published before 1963		published after 1963	
	no.	%	no.	%
(Ia) relation sociology/ social policy	12	13	1	2
(Ib) boundaries soc./other social sciences	6	6	—	—
(Ic) inter-relation theory and research	8	9	—	—
(Id) exact concepts	8	9	—	—
(Ie) methods and techniques	16	17	3	5
(IIa1) 'Grand Theory'	5	5	8	13
(IIa2) 'Middle Range Theory'	17	18	19	31
(IIb) theoretical-empirical	8	9	17	27
(IIc) empirical	—	—	6	10
others	13	14	8	13
total	93	100	62	101

published at the end of the 1950s and the beginning of the 1960s[1] were written by editors of the *Sociologische Gids*. Besides many of the editors were holding chairs in the newly founded subfaculties of sociology. Furthermore, the Table reveals that the elaboration of the programme was advanced to such an extent that in the period after 1963 more research results could be published: an increase of number of theoretical-empirical and empirical articles from 9 per cent (before 1963) to 37 per cent (after 1963). The data in this Table can also be used to say something about the 'colour' of the journal *Sociologische Gids*.

As the data suggest, some categories are clearly over-represented in the articles of the editors. In the period 1954–63 these are the categories: (a) the criteria for the relation between sociology and

[1] The textbooks I have in view are (titles translated in English): J. A. A. van Doorn and C. J. Lammers, *Modern Sociology, systematics and analysis* (1959); G. J. Kruijer, *Observing and Reasoning*, an introduction to the epistemology of sociology (1959); E. V. W. Vercruysse, *The Designing of a Sociological Investigation* (1960).

social policy (13 per cent), and (e) the criteria in methods and techniques (17 per cent). In the period after 1963 more articles have been written consisting of 'Middle Range Theories' (31 per cent) and articles in which middle range theories were the starting point for research (27 per cent).

As to the first category, Dutch sociology has always been very much oriented towards the application of sociological knowledge in social policy matters (this being connected with the sociographic tradition: van Doorn, 1964a; Peper, 1972). In the words of Horowitz, Dutch sociologists can better be classified as 'professionalists' than as 'occupationalists' (Horowitz, 1970, p.353). This designation fits especially well the sociologists affiliated with the *Sociologische Gids*. The emphasis on 'professionalism', according to Horowitz, is related to an emphasis on the testability of the propositions used in research reports. 'A premium is placed on reliability and on measurability.' (Horowitz, 1970, p.354.) In the programmatic phase, this accent is expressed by a high proportion of articles in the field of methods and techniques. Concerning the second period, the same emphasis is expressed in the publishing of articles consisting of good testable middle range theories and articles in which these theories are actually tested. In their investigation of Dutch Ph.D. dissertations, Lammers and Philipsen (1966) distinguished three modalities, viz. 'Parsonian', 'Mertonian', and 'Steinmetzian'. Regarding the articles written by the editors, the colour of the *Sociologische Gids* could be called in this terminology 'Mertonian'. Now that we have dealt with the general trends in the development of the *Sociologische Gids*, it is possible to look at the contribution to this development by the members of the core network. If our hypothesis, that the network concept provides a good representation of a scientific community is correct, then the nature of the contributions of the sociologist in this core network should be of a consistent type.

The collaboration among the members of the core network occurred before and after 1963. Stated otherwise, the network 'functioned' already in the programmatic phase in the development of the *Sociologische Gids* (although a small number of editors were admitted later to this inner group). Therefore, it is possible to investigate to what extent the development in the articles of all editors deviates from the development in the nature of the articles written by the core network. In my opinion, collaboration in scientific work presupposes shared conceptions of science. Although the analysis of articles in the journal is insufficient to provide a precise picture of the paradigm of a scientific community by comparison of the scientific activities of a group of scientists working together (in this case the core network) with those of other scientists (in this case the other editors of the *Sociologische Gids*), indications can be given with regard to the existence or non-existence of a paradigm in the field. This comparison will be conducted here by looking at the articles of the core network and those of the other members of the editorial board.

Table 8.4 *A comparison of the articles written by members of the core network with the articles of the other editors of the* 'Sociologische Gids'

	before 1963				after 1963			
	core netw.		others		core netw.		others	
	no.	%	no.	%	no.	%	no.	%
(Ia)	6	14	6	16	—	—	1	4
(Ib)	4	9	2	5	—	—	—	—
(Ic)	3	7	5	14	—	—	—	—
(Id)	6	14	2	5	—	—	—	—
(Ie)	8	19	8	22	2	7	1	4
(IIa1)	2	5	3	8	3	11	5	19
(IIa2)	9	21	8	22	12	43	7	27
(IIb)	5	12	3	8	10	36	7	27
(IIc)	—	—	—	—	1	4	5	19
	(total=43)		(total=37)		(total=28)		(total=26)	

It is apparent from Table 8.4 that, especially in the period 1963–71, there is a considerable difference between the nature of the articles of both categories of editors. While the difference between core network and other editors in the period before 1963 is relatively small, it is clear that in the last period the members of the core network published almost only articles in the categories 'Middle Range Theories' and 'theoretical-empirical'. The articles of the other editors are dispersed more equally over the four categories. This finding indicates that in the case of the core network one can speak of a relatively homogeneous group. Actually, the colour of the *Sociologische Gids* earlier indicated as 'Mertonian' can, for the second period at least, really be ascribed to this core network. The finding that in the second period the core network also published a good many research articles gives rise to the supposition that apparently not only agreement over theory and methodology existed, but that the paradigm was so elaborated that people had even come to the stage of 'problem solving'.

A more thorough investigation of the scientific conceptions of the members of this core network could give a more precise and complete picture of the scientific paradigm. In this investigation such a description was not possible for several reasons, one of which is the research strategy we have taken, tracing the development of a group of scientists' paradigm back to the social structure of their relationship. Future research into the paradigm should also pay attention to the relation of their paradigm and the 'subject matters' involved. For indeed, the sociologists in the core network are not working in the same fields in sociology, although, considering the articles, heavier emphasis is put on organization-sociology. This point, in particular, makes it difficult to answer questions concerning what problems are to be solved, in the opinion of the members of the core network. In conclusion, it can be stated that the latter was not really the purpose of this study, it rather intended to be an analysis of the process in a scientific community leading to the acquisition of a paradigm. Hopefully, I have succeeded in my purpose of clarifying some of the components of this process.

References

Barnes, J. A. (1969), 'Graph theory and social networks: a technical comment on connectedness and connectivity', *Sociology*, vol. 3, pp.215–32.

Bott, E. (1971), *Family and Social Network* (2nd ed.), London: Tavistock.

Crane, D. (1970a), 'Social structure in a group of scientists: a test of the "invisible college" hypothesis' in Reynolds, Larry T. and Reynolds, Janice M. (eds), *The Sociology of Sociology*, p.295, New York: David McKay.

Crane, D. (1970b), 'The gatekeepers of science: some factors affecting the selection of articles for scientific journals' in Reynolds, L. T. and Reynolds, J. M., op. cit., p.406.

Crane, D. (1972), *Invisible Colleges*, University of Chicago Press.

Curtis, J. E. and Petras, J. W. (1970), 'Community power, power studies and the sociology of knowledge', *Human Organization*, vol. 29, No. 3, pp. 204–18.

van Doorn, J. A. A. (1964a), *Beeld en Betekenis van de Nederlandse Sociologie*, Utrecht: Erven J. Bijleveld.

van Doorn, J. A. A. (1964b), 'De sociologische gids en de Nederlandse sociologie', *Sociologische Gids*, vol. XI, pp.20–51.

van Doorn, J. A. A. and Lammers, C. J. (1959), *Moderne Sociologie, systematiek en analyse*, Utrecht: Het Spectrum.

Hagstrom, W. O. (1965), *The Scientific Community*, New York: Basic Books.

Heyl, B. S. (1968), 'The Harvard "Pareto Circle"', *The Journal of the History of the Behavioural Sciences*, VI, No. 41, pp.316–34.
den Hollander, A. N. J. (1948), 'Sociografie en sociologie', *Geestelijk Nederland 1920–1940* (Proost, K. F. and Romein, J., eds) Deel II, Amsterdam/Antwerpen.
Horowitz, I. L. (1970), 'Mainliners and marginals: the human shape of sociological theory' in Reynolds and Reynolds, op.cit., p.340.
Kruijer, G. J. (1959), *Observeren en Redeneren*, Meppel: J. A. Boom.
Kuhn, T. S. (1962), *The Structure of Scientific Revolutions*, University of Chicago Press.
Kuhn, T. S. (1970), *The Structure of Scientific Revolutions*, 2nd ed. enlarged, University of Chicago Press.
Lammers, C. J. and Philipsen, H. (1966), 'Professionalization in Dutch sociology: an analysis of post-war doctoral dissertations and their authors', *American Sociologist*, vol. 1, No. 4, pp.197–203.
Lodahl, J. and Gordon, G. (1972), 'The structure of scientific fields and the functioning of university graduate departments', *American Sociological Review*, vol. 37, pp.57–72.
Masterman, M. (1970), 'The nature of a paradigm' in Lakatos, I. and Musgrave, Alan E. (eds), *Criticism and the Growth of Knowledge*, p.59, Cambridge University Press.
Mitchell, J. C. (1969), *Social Networks in Urban Situations*, Manchester University Press.
Mullins, N. C. (1968), 'The distribution of social and cultural properties in informal communication networks among biological scientists', *American Sociological Review* (33), pp.786–97.
Peper, B. (1972), *Vorming van Welzijnsbeleid*, Meppel: J. A. Boom en Zoon.
Phillips, D. L. (1972), 'Paradigms, falsification, and sociology', *Acta Sociologica* (forthcoming).
Popper, K. (1968), 'Epistemology without a knowing subject' in Rootselaar and Staal (eds), *Proceedings of the Third International Congress for Logic, Methodology and Philosophy of Science*, pp.333–72.
Reynolds, L. T., Vaughan, T. R., Reynolds, J. M. and Warshay, L. H. (1970), 'The "self" in symbolic interaction theory: an examination of the social sources of conceptual diversity' in Reynolds and Reynolds, op.cit., p.422.
Vercruysse, E. V. W. (1960), *Het ontwerpen van een sociologisch onderzoek*, Assen: Van Gorcum.
Westerdiep, A. R. (1970), *Werkkringen van sociologen*, (twee delen) Amsterdam/Groningen: SISWO/Sociologisch Institut Groningen.
Whitley, Richard D. (1970), 'The formal communication system of science: a study of the organization of British social science journals' in Halmos, P. (ed.), *The Sociology of Sociology*, Sociological Review Monograph No. 16, University of Keele.
Ziman, J. M. (1968), *Public Knowledge: The Social Dimension of Science*, Cambridge University Press.

Science, Scientists and Society

9

Elements from the debate on science in society: a study of Joseph Ben-David's theory*

Thorvald Gran

Introduction

Antonio Gramsci (1957) has suggested that the role of the scientist is not to bring science *ex novo* to the people, but to partake practically in the formation of systematic understanding of social inertia and social development. Everyone has some understanding. The scientific project is thus a continual transcending of old and presently existing modes of production and modes of thinking. Or in other words, the project is to make good sense out of common sense. Therefore science should be submerged in common man's striving for a better life, not floating above it.

A somewhat different view is the position that science comes to society from the outside. Even if specified in different ways this view seems to be the prevalent one in much Western sociology of science, R. Merton (1970), M. Polanyi (1951) and Ben-David perhaps being the major representatives. The core of this position is that the methods and contents of scientific thought essentially develop according to their own, immanent logic. Society can further or hamper the development of science and more or less eagerly apply its products, but it is in the minds of working scientists that we can and must find the basic logic of the scientific enterprise.

* I am indebted to Ben-David and colleagues at the ISA conference on the sociology of science in London, and colleagues at the Institute of Sociology and Institute of Philosophy in Bergen for helpful comments on drafts of this chapter. Grants from the Norwegian research council, NAVF, have made this work and participation at the London conference possible.

However, it is important to note that the viewpoint of 'science from the outside' can be examined to accord with Gramsci's analysis. In the book mentioned above (1957, p.91) he points out how dominant intellectual systems can influence the masses as an external political force, limiting the original thought of the popular masses negatively while at the same time binding the leading classes together. Here 'outside' is related to the class conflict. By the above mentioned sociologists it is related to society as a whole. Whether the position that scientific work is essentially socially neutral is common sense I shall not say. It at least seems widespread in intellectual circles, thus making a study of the position worthwhile. I will risk taking Joseph Ben-David's work *The Scientist's Role in Society* (1971) as representative of this position and submit the main theoretical points in it to critical analysis.[1] This being the project, it is first necessary to pinpoint some general Marxist positions on the relation of society to scientific production.

Marx presented a devastating critique of German idealism without ending up in its antithesis of economic determinism. Programmatically he states how the socialist economic principle is one side of human reality.[2] The other side, and thus equally important as object of study, is man's theoretical reality (religion, science, etc.). And he points out that science is not the presentation of a set of alternative dogmas to replace the old ones but the development of new forms of social life through a systematic critique of existing forms and the contradictions they contain. Marx points out that the technical divisions of work and the social division of work between classes became fully developed when the degree of surplus production in society made complete separation of material and theoretical production possible (cf. Marx, 1970, pp.56–122). This institutional isolation of theoretical work made 'unreal' thinking possible—made it possible, for example, to comprehend the interests of the established bourgeoisie as the interests of the whole society.

Following Althusser's formulation (1969), we can thus say that elements in the superstructure of society, the production of knowledge and understanding being one of these elements, have a relative autonomy—the degree of autonomy in the last instance determined by the

[1] We will in this chapter abbreviate J. Ben-David as B-D. Interpolations of a semantic character in quotations from B-D's work will be found between /and/.

[2] In a letter to Arnold Ruge made public in *Deutsch-Französische Jahrbucher*, Paris, 1844, published in the Norwegian Marx-translations *Verker i utvalg*, No. 1, Pax Forlag, Oslo, 1970, p.51.

class relations in the existing mode of production. In this way science can be an important driving force in the reproduction and/or revolution of a social formation, but with the capacity of the productive forces and the class relations in the economy giving the autonomous structure its character and setting its limits.[1] That the substantive structure of scientific thinking springs out of science itself—that the qualitative aspects of science so to speak, are in some abstract sense autonomous so that society only influences the quantitative aspects of the scientific enterprise, is alien to this understanding. The dominant ways of thinking have a relative autonomy, but they cannot be independent of, or in deep contradiction to, the interests of the ruling classes.

In other words, we have to distinguish between unmediated relationships between class and science and relationships that are mediated through complex sets of socio-economic structures. It is in this latter theoretical setting that the problem of the class character of positivist science is located. In the Marxist perspective, then, scientific categories are theoretical expressions for the development of the relations of production—under capitalism expressions for the antagonistic relation between labour and capital. Scientific work, like all other aspects of production (production seen as social interaction mediated through the material structures), is dependent upon the production relations and active in their change and revolution. It can lift itself out of this process to proclaim its basic neutrality only with ruling class support and as an ideological act (proclaiming that the scientific method is above ideological battles and that results are not true or false, only not yet falsified approximations to the given laws of nature[2]).

Marx criticized Proudhon for viewing science and its categories as something produced autonomously by scientists. He states:

> As soon as the historical development of the production relations are left out—and the scientific categories are theoretical expressions for these relations; as soon as one views these categories as ideas that have been developed autonomously, then, whether one likes

[1] We must on these grounds refute the position of B. Barber (1952, p.30) when he states: 'The burden of the Marxian view on these matters is that science is a wholly dependent part of society, molded fundamentally by the economic factor; and that therefore there is no reciprocal influence between science and the other components of society.'

[2] Nature here denotes 'man in nature'. Later we will see that B-D uses the term 'laws of nature' in the natural science sense of 'laws independent of man in physical nature'.

it or not, one must move the origin of these ideas to the movement of pure reason.[1]

The problem posed

As work and thought form a unity at the societal level a transcendence of the class divisions is part and parcel of the development of a science working for the emancipation, rather than the manipulative conservation, of class structures. Typical of manipulative (technocratic) science is exclusive interest in means-end relationships and prognostication. How we reach given ends and the probability of reaching them under different conditions is a central focus. This excludes the study of the nature or character of social systems and of how quantitative changes contain qualitatively new structures. Such study is quickly termed ideological or dogmatic (cf. Kolakowski, 1972).

For technocrats the deepest nature of things are given. We shall see that for B-D the nature of science is unproblematical. It has basically always been the same thing. He points out that the attempts that have been made at studying the nature of science have been unsuccessful because they have not made the prediction of knowledge possible (B-D, 1971, p.8). Therefore B-D overlooks the problem of the changing nature of science and limits his study to the emergence of the role and the organization of the scientific community. His problem is basically: At what time and where were the conditions favourable to the establishment of a vigorous scientific community? Who had the *means* to realise this *end*?

But B-D does not discard the problem of the class basis of science offhand. He uses the example of the rising merchants' relationship to liberal ideology to argue the refutation. The argument is as follows. In liberalism, as in science, individualism and rationalism are taken as important elements. Now if the interests of the rising merchant classes were constituent parts of science then the merchants should have actively favoured liberalism as an ideology. But in the writings of Hobbes and Adam Smith, B-D finds that the merchants, and liberal ideologues, often supported absolute monarchy and monopolistic positions in the economy. This, according to B-D, should cast serious doubt on the positive relationship between class interest and the nature of empirical science. The merchants were not in support of any ideology, short-range maximization of profit was their basic interest.

[1] My translation from the Norwegian text in Marx (1970, pp.124–5).

Here B-D seems to accept that there is a relationship between the bourgeois class and empiricist science only if the relationship is a direct manifestation in thought and in practice. Now if we view the relationship as a mediated relationship, and distinguish between ideology and actual economic practice, we can gain a somewhat different understanding. That the merchants actually practised monopoly does not then prove that the merchants, as a class up against church and aristocracy, were not systematic supporters of liberal ideology and thus of empirical science as B-D defines it. The ideology of the bourgeoisie has often stated that they are quite free from any ideology, that society under capitalism is a free for all (liberalism), and that rational calculation (empirical science) is the way to solve disputes and develop society. That positivist science actually aids in veiling the class character of capitalist society and at the same time actively furthers the development of new products and machinery is, of course, the practical side of the matter.

Central concepts

In this section I will briefly present and characterize the central concepts in B-D's study. Some of the problems they raise will be looked into under subsequent headings. The central concepts are 'science', 'role' and 'institutional approach'. First, his definition of science which we can infer from his description of what scientists do (1971, p.1):

> scientists are engaged in discovering 'laws of nature' that cannot be changed by human action. Thus not only are they faced—as in mathematics—with the immanent logic of their own system of thought, but they accept the further restriction that their systems have to fit the structure of natural events. In principle, this is true also of social scientists and scientific students of culture.

There are three important elements in this definition. (1) Scientists are fundamentally in search of a positively given set of natural and social relationships (laws). (2) These existing relationships are independent of human action. (3) The fundamental logic of man's thinking corresponds to these basic regularities in nature. These elements together suggest that the scientific project is to work towards an ever better matching of thought to reality. This definition suggests an empirical/analytical understanding of science. We will, in the next section, try to explicate the contents of this category and some of the theoretical consequences of using it.

B-D chooses the sociological concept 'role' as his working concept. A role 'is the pattern of behaviours, sentiments, and motives *conceived by people* as a unit of social interaction with a distinct function of its own and *considered* as appropriate in given situations' (pp.16–17, my italics). This definition we suggest is thoroughly rooted in the idealist tradition of modern Western sociology. We will, in the next section, try to clarify the close relationship between formal or analytical empiricism and the idealist understanding of society.

The last of B-D's central concepts that we must analyse is what he calls the institutional approach. Rather than the study of interpersonal interaction this approach implies a study of the social/structural conditions for the development of organized science. One set of these conditions, which from a materialist point of view is important, the economic conditions, is summarily defined as 'level of wealth' and discarded. The reason for this is that studies show that the correlation between wealth and scientific contribution is low. Therefore 'the economic conditions can be regarded as given' (p.14). From this short visit into the materialist realm of study B-D quickly retreats to idealist concepts (p.16, my italics):

> More probably, the two economic conditions and scientific growth are related by a common underlying characteristic, such as *talent*, and the *social motivation of advancement*, or something similar.

What we here can say of these concepts is that they support the common sense understanding of all science as natural science, science as not only institutionally distinct, but also functionally exterior to other forms of work in society and lastly that changes of motives and values rather than changing class conflicts and their material base, are the driving force in the historical transformation of society.

The paradigm

Alain Badiou (1970) characterizes as bourgeois epistemology the distinction, and the unity, between the concepts of empiricism and formalism. In this view, science is the process of making ever better fits between the objective given reality and formal theory. R. Carnap gave explicit form to the distinction in the late thirties while W. O. Quine suggested that the distinction was unnecessary: being and being a value on a variable were identical. Thus science is primarily the project of creating a formal (analytical) representation of the given world. At the limit of bourgeois epistemology, Badiou suggests, is the project of constructing the model of how the model-constructors themselves (the scientists) function. Thus the structure of the human mind becomes some sort of supreme model and mathematics the supreme language because it is supposed to be directly related to human logic, without 'errors', without substantive contents.

B-D's understanding of science clearly falls within this empirical/analytical 'model'. A necessary counterpart to this model is empirical scientific speculation, (cf. Poulantzas, 1969). Stated briefly: The more imagination you (as a scientist) have for constructing concepts and models the greater is your chance to contribute positively to the refinement of the scientific description of the objectively given and basically invariant world. B-D adopts this view of science when explaining why the development of concepts, methods and/or theories of science had nothing to do with class interests (p.10). In his view science has been autonomous since its emergence. If there has been a relation between science and class interests it has been an external relation such that scientists could choose between competing theories and ideologies, with the purpose of letting the objective facts decide what actually existed. Thus (p.11, my italics):

> Ideological *bias* (socially determined or not) might have played some role in the *blind alleys* entered by science, /but/ the philosophical assumptions that had become part of the living tradition of science were *selected by scientists* from the array of competing philosophies for their usefulness in the solution of specific problems and not for any social determined perspective or motive.

It is, of course, not surprising that this deification of the thinking of the scientists leads to a scientific focus on thinking about the world in general. Such an understanding, represented by the analysis of ideas against ideas (which of necessity leads to a focus on individuals), can be termed idealism. The idealist approach can best be seen in the definition of theoretical concepts. Let us look at some of these.

In bourgeois sociology concepts like status (*perceived* place in the social hierarchy), authority (uncritical *acceptance* of orders/information) and roles (other people's *expectations*) are of central importance. Now only individuals can think and feel. Groups or social classes cannot. Thus groups and classes are seen as sums of individuals who think alike (cf. Aubert, 1964, p.115). Marxist sociology does not *en bloc* discard these two positions (empirical formalism and idealism), but transcends them in important respects. Referring back to Marx's and Gramsci's positions on science we see that science is not primarily the creation of a plausible image, a formal representation of a given world, but the project of a practical transformation of society. The Norwegian social scientist R. Enerstvedt (1971, p.90) states it thus:

> It is the character of marxism not to be dogmatic. Marxism is not a closed philosophical system, but a way of looking at the world where change—not interpretation—is the highest principle.

In other terms we could say that the scientific project is the transcendence of limitations on social, and thus on individual, life. In this way the purpose of science goes beyond prediction and moves actively into the realm of aiding the realization of possibilities that are 'held back' by either natural or social forces. A central mechanism of domination is to present man-made social structures as inevitable natural structures, or, not to present them at all. Showing the actual changeability of such 'quasi-natural' structures is thus a scientific project.

Now taking what people positively and individually think as the basic elements of reality of course threatens the project which has been defined above. So does the focus on individuals because as individuals we are virtually powerless. Social power is, in the Marxist view, located in social classes and social classes are not primarily groups who think alike but groups standing in the same position in the productive process. By focusing primarily on the modes of production and the class contradictions in these modes the idealist position is potentially transcended. This approach does not give any simply and unambiguous definition of class, but it does give a concept with a theoretical status quite independent of individuals' values and preferences thus making, among other things, analysis of ideological hegemony possible.

These comments set B-D's definition of science in a new context. If it is correct that class structure and the structure of science are internally related through complex practical and theoretical mediations and that antagonistic class contradictions actually exist in the societies that B-D has studied, then the empirical formalism of B-D is a definite position in a conflict. But when such a position is transformed to a class neutral and universally applicable definition, the opposing positions obviously are eliminated from the study. Equally obvious, if such a definition is accepted as common sense, an actual conflict is transformed to an harmonious phenomenon of nature. Thus we see in B-D's study that the definition of science adopted by the medieval Catholic church just plainly was not science. The same probably applies to the Marxist position on science as it is explicated above.

Analysis

It is not difficult to see the affinity between B-D's definition of science, his theoretical paradigm and the choice of an evolutionary market-model as his substantive theoretical framework. I can only indicate the connections here: (1) Science has, by definition, always been qualitatively the same project: thus evolution. (2) Science is seen as being 'above' society, thus a central focus on the community of scientists as the developers of science. (3) His definition of 'role' points both to the individualism (only individuals have ideas and motives) and to the universally valid 'price mechanism' (persons and ideas are accepted into the scientific community when they are commonly evaluated as appropriate) typical of the market model. This model can certainly be useful in studying the maintenance and spread of empirical/analytical science, but B-D's whole approach excludes the possibility of analysing the emergence of empirical formalism as a specific form of science and its rise to a position of domination or hegemony.

(a) *Diffusion of science*

The development of empirical science is seen by B-D, in both time and space, as a process of diffusion from centre to periphery; the invariant and universal norms of science deciding who gets into the community and the varying social conditions determining the location of the centre and how quickly science gets established. Science thrives when the structure of society 'matches' the structure of science. Thus we hear about the English case (p.77):

> Finally starting from the 1660s there emerged a series of attempts to shape political and economic philosophy and practice according to the model of self-regulating mechanical systems/i.e. the scientific model/, rather than as an order imposed by supreme authority. Thus political society was conceived as composed of independent individuals, as matter was composed of atoms, and as held in balance by . . . conflicting forces . . . .

Or in the form of generalized theory (p.170):

> Between the fifteenth and the seventeenth centuries there arose influential groups . . . in different places in Europe who were in search of a cognitive structure consistent with their interest in a changing, pluralistic and future-oriented society. Empirical natural science (whose conceptual development was quite independent of these social circumstances) provided such a cognitive structure of testable validity.

Thus we are told that science proper develops through diffusion in, and expansion of, the scientific community, but that adoption or institutionalization of science is dependent upon the formation of group interests in the different societies. This view is applied to the country-specific studies in the book. But on the level of general theory the diffusionist model is also applied to the internal organization of the scientific community (p.171):

> Having established the general evolutionary pattern of scientific growth, our study has to show what the mechanism of selection of a certain type of role and organization was. It appears that this mechanism was competition between strong units of research operating in a decentralized common market for researchers, students, and cultural products.

This seems to be a relatively good example of how B-D's model of autonomous science can function as a straight-jacket. The social conditions outside the unitary (but disturbable) international scientific community that are introduced in the empirical studies recede into the background the more general the theory becomes. At its most general level both the role and organization of the scientific community are functions of competition and needs within the scientific community.

(b) *Science alien to practical concerns*

A fundamental thesis in B-D's study is that science cannot thrive where practical interests dominate. Thus practical engagement in the betterment of man's physical and social conditions was the main reason for the lack of institutionalized science in traditional societies (pp.23–32). The short spurts of scientific creativity in these societies depended on the rise of conflicting ideologies, thus creating a market-situation of ideological pluralism. Once these conflicts receded and ideology became unitary again, practical interests got the upper hand and science deteriorated.

How then did these practitioner-scientists come to view themselves as pure scientists pursuing science for its own sake? B-D discusses this problem at great length and I want to suggest that it is here that the analysis most clearly exposes its idealist tendency. Let me try to pinpoint his argument.

(1) It is decisive for the existence of the scientific role that the scientist is primarily engaged in science for its own sake. (That is, interested in explaining 'nature analytically in order to predict natural events', p.29.) B-D explicitly presents this argument (p.45):

> ancient science failed to develop not because of its immanent shortcomings, but because those who did scientific work did not see themselves as scientists.

(2) However, the scientist's personal view of himself as a pure scientist is not a sufficient condition for the establishment of the scientific role. Equally necessary is that the role is commonly accepted, that the 'social function of science' is accepted in society. Thus B-D says (p.39) about Aristotle's Lyceum, where real scientific embryos developed:

> Had this philosophical purpose /science for its own sake/ been accepted as a social value, and had it been consistent with a genuine freedom of disciplines, then a scientific role could have arisen that would have been recognised as a social function.

Here B-D seems to imply that if the Greek thinkers in the Lyceum had chosen science for its own sake, and the scientific role, as B-D defines it, had been accepted in society at large, then science could have been 'recognised as a social function' already at that time. Besides raising the problem of how something can be 'accepted as a social value' without being 'recognised as a social function', the implications of such a view of development are obviously idealistic.

Now why did the role become established in seventeenth century Europe? B-D suggests several reasons, but essentially he argues as follows. At that time there was wide agreement on the superiority of the experimental method. The reason for this agreement was the fact that the method was above the warring ideologies (p.73):

> Without an agreement on the experimental method, however, an

> autonomous scientific community could never have arisen. Had science been presented as a superior, but logically closed and coherent philosophy, it would have become one of the warring philosophies rather than a neutral meeting ground.

The pressing problem here is what was the social base of this non-ideological method of getting knowledge? It could not emanate from the scientific community because it was by definition not in existence. Furthermore, it could not be a method of knowledge-production specific to the rising bourgeois classes because the method, again by definition, was above specific class positions in society. So where did it come from?

The point made by B-D that it already existed in embryo in the Lyceum may give us a clue. Looking back at B-D's definition of science, the explanation may be the following. Pure science as *idea* has *always* existed. What until the seventeenth century Europe was missing were the conditions favourable to the institutionalization of the idea in society. This of course conforms to the idealist interpretation. It makes the question of the origins of pure science meaningless, leaving the question of conditions or means for the realisation of this set of ideas as the one to be studied.

(c) *The succession of scientific centres*

The focus of B-D's study on science in Italy, England, France, Germany and USA is thus the following question: When, and under what social conditions, did 'science for its own sake' become institutionalized? The theory which B-D presents on this point, and which as far as I can see is used to explain both the establishment and the diffusion of science, can be summed up as follows.

The basic condition for a vigorous science is a strong 'scientistic movement' in society, that is a group of people (mostly non-scientists) who believe in science as a valid way to the mastery of nature and to the solution of social problems (science as a symbol of the infinite perfectibility of the world, p.78). The relative strength of this movement in different countries decides which country will be 'the centre'. On the other hand the degree of 'institutionalization of science' is the decisive factor explaining the decline of science. Institutionalization often leads to a decline in the scientistic movement and a tendency for scientists proper to become more and more engaged in practical affairs.

This theory of science has an interesting dialectical structure which can be compared to the Marxist theories of class conflicts and the state. The rising classes (the bourgeoisie in relation to the different formations of feudal classes, the proletariat in relation to the bourgeoisie at later stages) fought for its own, but at the same time for the progressive interests of the whole society in attacking ossified forms of minority rule. But through the stage of institutionalization (i.e. through the formation of a new state apparatus after ousting the established ruling class) the earlier rising class itself ossified, at least as long as antagonistic class contradictions continue to exist in the new situation. Thus the process continues.

The undialectical aspect of B-D's theory is contained in the view that science from the earliest times has always been the same. The scientistic movement has, so to speak, always been the qualitatively same movement. It has only quantitatively and cumulatively gained strength, the dialectic between movement and institutionalization principally deciding the location of the centre of science, not affecting the nature of science. This harmonizes well with the view that ossification and engagement in practical affairs go hand in hand. This theory is applied to the countries mentioned, and B-D presents a host of interesting facts on the work of the classic scientists and the institutional battles that were fought. It is, however, outside the scope and purpose of this chapter to sum up this analysis here.

Politicization and scientific pluralism

Let us try to draw the threads together. We can perhaps depict B-D's essential argument in the following models of the growth and decline of science.

	Growth	*Decline*
Social base	Scientistic movement	Extra-scientific interference
Substantive structure	Empirical analysis (based on objective physical phenomena)	Metaphysical speculation (based on subjective 'world views')
Problem focus	Contemporary practical problems[1]	Historical disciplinary problems
Socio-political tendency	Liberal pluralistic social reform	Authoritarian political potential

[1] When B-D points out that focus on practical rather than disciplinary problems is a growth factor he seems to mean that this focus must be freely chosen by the scientist.

We have tried to raise questions concerning the intrinsic theoretical and ideological contents of B-D's definition of science. It provides an understanding of science as an autonomous, cumulative and unitary process. While not denying that these are important aspects of science we have argued for the necessity of studying the internally mediated relationships between the changing structure of science, the institutionalization of science and the 'scientific interests' of the changing formations of ruling and suppressed classes. With B-D's definition of science the problem of epistemological breaks, or changes of paradigm, the problem of the periodization of science and the synchronic, internal contradictions in the scientific enterprise, e.g. between analytical empiricism and dialectical materialism, seem to disappear.

B-D's analysis is perhaps present in its most pregnant form in the problem of the politicization of science. Politicized science seems to be the direct contradiction of B-D's concept of autonomous science. His exposition reaches dramatic heights with this problem on the agenda (p.180):

> If science is perceived as partial to some social interests, and scientists are seen in an invidious light, then people may start doubting the moral value of seeking scientific truth for its own sake and applying it for the purpose of changing the world. This may spell the end of scientific culture.

Need the politicization of science spell the end of scientific culture? If the direct opposite of B-D's position on science is adopted we definitely are in danger. This position would be: the scientific enterprise is in all aspects and forms directly in the hands of a ruling class.

However, if in fact we take the view that basically science is not one unitary thing and that the formation of scientific methods and theoretical structures are, in important aspects, intrinsic parts of the class struggle, two positive consequences follow. One in relation to B-D's conception of science, and the other clearly showing the value of what we call the pluralistic university.

First, the consequence of this view on B-D's conception of science would be to open the closed aspects of his definition. The idea that there is actually one set of laws, quite exterior to all social classes, ruling basic parts or structures of our social lives would be broken. The idea that the development of the substantial structure of science

---

It is when extra-scientific institutions press their problems on the scientific communities that the change from growth to decline can set in.

is something produced autonomously and cumulatively by the scientific community proper would also be opened for critique and development. While the danger of the antithesis to B-D's position is direct bureaucratic control of science, the danger of B-D's thesis itself is a combination of perceived purity in the scientific community with, in practice, a subordination of science to the interests of the different bourgeois classes dominant in Western Europe and the USA.

The second consequence concerns the pluralistic university.

(1) Recognizing that science is an integral but distinct part of both bourgeois and proletarian classes implies, of course, a recognition of both the value and the necessity of distinct institutions for specific scientific production.

(2) The idea of a pluralist university is, in principle and in practice, of central value for scientific development. In principle the idea of diversified and democratic science is of central importance to socialism. In practice pluralism means that once the abstract 'above-society' definition of science has been discarded an open confrontation between the political/scientific groups and the functional groups within the universities can take place. Although B-D's definition of science gives the concept of scientific pluralism a meaning alien to socialism, the reality of the pluralist university in this sense is necessary for a democratic development of the scientific enterprise today.

References

Althusser, L. (1969), 'Motsigelse og overbestemmelse' in his *For Marx*, Oslo: Förlaget Ny Dag. The original was published in Paris by F. Maspero in 1965 and by Allen Lane in London in 1969.

Aubert, V. (1964), *Sosiologi*, Oslo: Universitetsförlaget.

Badiou, A. (1970), *Le Concept de modele*, Paris: Maspero.

Barber, B. (1952), *Science and the Social Order*, Chicago: Free Press.

Ben-David, J. (1971), *The Scientist's Role in Society: A Comparative Study*, Englewood Cliffs, N.J.: Prentice-Hall.

Enerstvedt, R. (1971), *Den Intellektuelles Dilemma*, Oslo: Förlaget Ny Dag.

Gramsci, A. (1957), *The Modern Prince and Other Writings*, New York: International Publishers.

Kolakowski, L. (1972), *Positivist Philosophy*, Harmondsworth: Penguin.

Marx, K. (1970), *Verker i Utvalg*, 2 vols, Oslo: Pax Forlag.

Merton, Robert K. (1970), *Science, Technology and Society in Seventeenth Century England*, New York: Fertig.

Polanyi, M. (1951), *The Logic of Liberty*, London: Routledge & Kegan Paul.

Poulantzas, N. (1969), 'Political ideology and scientific research' in Dencik, L. (ed.), *Scientific Research and Politics*, Sweden: Lund.

10

Societal influences on the choice of research topics of biologists

Albert Mok and Anne Westerdiep

Politics, science and the sociology of science

Science, at least that part of the scientific endeavour which is called pure science, is generally viewed as apolitical, and scientists are mostly seen as people who go about their business and leave politics to the leaders (Greenberg, 1969, p.31). These views are reflected in the sociological literature: very little research has been done on the political engagement of scientists or on the scientists' concern with societal problems.

Neglect of the political aspects of the scientists' work is part of what Whitley (1972, p.63) has called the ideology of 'black boxism': sociologists of science are almost exclusively concerned with the visible inputs and outputs of the scientific system and are paying very little attention to the 'translucid box' kind of sociology of science, the study of the process of science itself. Sociologists of science, says Whitley, are preoccupied with the producers in a way that takes little account of what is being produced, why it is being produced, or what it is being produced for.

In this chapter, we would like to contribute to the 'translucid box' kind of sociology of science by discussing briefly an exploratory research project into the selection of topics by biologists working mainly in ecological laboratories. We give our experiences with research of this kind which, not surprisingly, proved rather difficult. In general, it is questionable whether the survey method which we applied, including scaling, is appropriate for research of this kind

(Lemaine *et al.*, 1972, p.12). In this case though, the applied method had a special meaning, as will be seen later.

Another problem is whether the environmental situation in the Netherlands is as specific in producing societal pressure on biologists as we claim it is. Is the Netherlands a special case? Would a similar environmental situation have similar effects in other countries? We do not know, but at least these are interesting questions to ask. Much comparative research will have to be done to answer them.

Our starting point was that, as biologists become politically oriented, or 'concerned', they tend to derive their research topics from external sources. On a behavioural level, one could say that as scientists become politically oriented, that is show concern and responsibility for the application of their products to benefit society as a whole, they not only tend to take part in societal action, but also tend to view their science itself as a means to reach these larger societal goals.

However, as our project progressed, the structural conditions of the research institutes where biologists work proved to be of the utmost importance. We, then, developed the hypothesis that taking part in societal action may be a response to an inflexible laboratory situation and lack of career opportunities for biologists in 'pure' research. Though we may not have collected sufficient data on science policy to 'prove' anything in this respect, we can at least give some indications for future sociological research. That will be done at the end of this essay.

In the Netherlands, as perhaps elsewhere, science policy favours long-term research projects, and sponsors demand extensive programming of research projects as a means of controlling expenditure and utility. Both these factors tend to make for inflexibility in the choice of research topics by the laboratory management and a stable employment situation.[1] Formerly, biologists reacted with resignation to their situation if they were at all concerned. They did their research on bark or ferns without linking them to man-made material objects. Now that biologists are awakening, as it were, to the environmental situation, long-term research programmes, some of them established in a period with little environmental concern, come to be felt as a restraining force by those biologists who are action-oriented, who

[1] It should be noted that this stable employment situation is at least partly due to the fact that scientists working in government or government-subsidized research institutes (like all university laboratories) have tenure.

want to choose topics because of their relevance to society and who want to keep control of the uses which are made of the results of their research. Short of leaving pure research altogether, they will have to realise their political orientation within the limits prescribed by existing research programming, in small short-term projects. Bigger, long-term projects, which are often sponsored by basic research funds, have to be approved by laboratory heads and project leaders and, if realised in team work, by colleagues. Judging from the research done in France by Lemaine and his collaborators (Lemaine *et al.*, 1972, p. 146), the intellectual climate of biological laboratories, as indicated by the reputation of their heads, is bound to be an important factor in granting autonomy to researchers to go outside of the established research paths.

Although much research has been done on if and how research climate favours or restrains intellectual productivity (Pelz and Andrews, 1966), we are not so well informed about the reactions of scientists to external pressures. That is what the next section is about.

Sources of focuses of research interest

Roughly two main sources of focuses of research interest can be distinguished:

(1) internal sources, those sources which are endogenous to the scientific community and the research establishment;

(2) external sources, those sources which are exogenous to the scientific community and the research establishment.

Zuckerman and Merton (1972) have drawn attention to the fact that most authors stress the internal, institutional source of change, to the neglect of external ones, 'those that come from the environing society, culture, economy, and polity' (Zuckerman and Merton, 1972, p.346). They do not operationalize those external factors, but rightly criticize Kuhn (1970) for being too restrictive in saying that the sociological form of the answers to questions of this kind must ultimately be in terms of a value system and the institutions that transmit and enforce it. Kuhn's work being influential as it is, it is small wonder that many sociologists of science follow his lead in picturing the scientific community as an essentially closed society (Watkins, 1970, p.26), working in 'self-imposed isolation' (Hancock, 1971, p. 169) and jealously guarding its independence and integrity. Much of the literature in the sociology of science is devoted to analysing the mechanisms by which the scientific community, no matter whether it is seen as an institution or as an interaction structure (Ben-David, 1971; Crane, 1972), brings about conformity in science, not in the least about the problem areas to be investigated. Mulkay (1972), to take a recent example, describes the research community as a complex web of problem networks, encouraging certain lines of work while discouraging other ones.

All these authors stress the effects on the scientist of socialization into the scientific community: getting to know the norms governing scientific activity, learning to value the reward system of science (Hagstrom, 1965) and letting the choice of research problems be governed by purely internal criteria, like the advancement of the field and the career of the scientist (Ravetz, 1971). Apparently, the scientist is taught to claim the freedom to choose his own research problems, but at the same time not to go outside the limits prescribed by his peers. It seems to us that the factors internal to the 'system of science' insufficiently explain important changes (revolutionary or not) in focuses of research interests which have recently occurred among scientists engaged in pure research in the biosciences.

In times of crisis, when pure science comes under attack and is challenged by society to solve its 'real' problems, it is quite natural that scientists should turn to the outside world for research topics. In those circumstances, scientists who go on thinking that 'the external world is not something to be concerned about, but rather a passive object to be analysed', to quote Ravetz (1971, p.430), will feel greatly strained. Then, it may well be that the professed norms of the scientific community prove incapable of providing real guidance for action (Barnes and Dolby, 1970). Perhaps, we should not, as Ravetz suggests (1971, p.415), equate finding the source of research topics in the outside world with the feeling of moral responsibility for the effects of this work. But we do think that these two factors are closely connected. Scientists professing the norms of the scientific community to the exclusion of other guidelines for behaviour will have great difficulty in acting in direct interaction with society's needs. Ravetz, it seems to us, is right in claiming that those scientists deriving their research problems from exogenous sources are likely to be more 'concerned' than those scientists who derive their focuses of research interests from some internal source. But we are not so sure about the causal relationship. Do 'critical' scientists find their topics within the area of concern *because* they are 'concerned'? And are critical scientists who do not do research within the area of their concern necessarily less concerned? It seems a plausible supposition that the latter scientists will want to reduce the dissonance between the norms of the scientific community and their own feeling of moral responsibility for the effects of their research by turning to activities outside their direct work sphere.

In this brief research note, we examine how a general increase in ecological awareness among Dutch biologists is related to choice of research topics and involvement in environmental movements. It will be remembered that this is a preliminary analysis of the data and our intention is now to raise questions and not to claim conclusive results. It, also, remains to be seen whether the Netherlands is a special case. If not, our findings should not be generalized before having established whether similar circumstances in other settings have similar effects on attitudes and behaviour of biological researchers.

Biologists and the environmental situation in the Netherlands

We have some indications for our hypothesis that Dutch biologists are becoming more and more 'critical', or politically oriented, that they tend to see their science less as a goal in itself and that they show this attitude by taking part in societal action. Concern about the future of mankind (Handler, 1970) has stimulated a number of biologists to adopt the role of 'Doomsday Scientist', although it is true that many researchers even in the biosciences have stayed aloof and do not wish to be concerned (Chisholm, 1972, p.149). Years ago, concerned biologists began warning society that a number of developments in agriculture and industry were threatening our capacity for survival. Only recently have their books been appreciated and some of these became bestsellers (Carson, 1968; Commoner, 1968, 1971; Ehrlich, 1968; Taylor, 1968). The environmental situation is a matter of concern in all industrialized nations, but we have reason to believe that this is especially the case in the Netherlands.

Owing to its geographical position, there is a unique combination of biotic, abiotic and psychogenetic environmental factors in the Netherlands. The country has a climate without extreme differences in temperatures. There is a constant supply of water to an already considerable amount of water. The soil is a peculiar combination of quartzy sand and very fertile clay, almost without rocks. And all this is situated in a relief that is perhaps monotonous from the point of view of scenery, but which sets almost no limits to human activity.

These combined circumstances account for the fact that during the last twenty-five years almost without notice the country's resources have been gradually poisoned by the human uses of the environment. At the end of the 1960s the Dutch people suddenly became aware of their environmental situation. Proof of this heightened awareness can perhaps be found in the following facts. In a survey about the expectations of the future held in 1970, it was found that environmental concern had replaced the housing issue, which had been people's concern ever since 1945 as the number one problem to be solved (van Dieren, 1972). Six months after publication, more copies of the MIT report on the limits to growth were sold in the Netherlands than anywhere else in the world. In 1971, we counted no less than 105 major articles on environmental issues in two national dailies and three national weeklies. In March 1972, a permanent information centre for the biosciences was founded in Utrecht, the geographical heart of the country, where any concerned citizen can get expert information about the effects on the environment of biological, chemical and medical research.

The Netherlands is the most densely populated country in the world. If the whole of the world population (3½ billion people) lived in the United States, that country would still not reach the population density of the Netherlands. This means that pollution problems are much more acute than anywhere else in the industrialized world. Two of the major rivers of Western Europe, the Rhine and the Meuse, have their outlet in Holland, so that much of the industrial pollution from France, Belgium and Germany finds its way to the Dutch Delta and to the North Sea. After the 1953 flood disaster, the decision was taken to close the Delta, and so gigantic works to this effect have been in progress since 1954. But only very recently have people started to realize the disastrous consequences for the biological equilibrium of the closure of waterways and of the founding of ever new polders. The ideology of the reclamation of new land for agricultural uses, long used to justify the expenditure of billions of guilders, now seems to have reached its end. Because of the limits to the available space, planners are confronted with the situation that solving environmental problems in one area means creating problems in others. These and many other problems have created the basis for the institutionalization of the concern of the citizens. Action groups have sprung up everywhere, and many permanent environmental associations have been founded by people who want to prevent the environmental disasters which they envisage. Most of the actions have been directed towards a concrete object, like preventing the closure of one of the major waterways of the Delta (the Oosterschelde) or preventing the islands in the North from being united with the mainland by reclaiming the land in between (the Waddenzee). In all cases the principal argument is the drastic change in the biosystem as a result of the human effort. In this situation, laymen have turned to scientists, especially to biologists, for help and guidance, and for solutions for some of the gravest problems. Biologists are likely to react in two ways:

(1) by staying aloof and by hiding behind a cloud of objectivity and neutrality; this is evidenced by a withdrawal to the norms of science (Merton, 1957), especially the norm of scepticism; these scientists say that 'we do not know enough to find solutions for these problems';

(2) by actively engaging themselves in all matters of concern to citizens with regard to the environment.

These latter 'concerned' biologists especially are likely to undergo exogenous influences on the choice of their research topics. These should be evidenced by the following activities:

(1) a desire to do ecological (as opposed to general) research projects in the laboratory;
(2) a tendency to hold lectures and write articles for lay publics;
(3) participation in environmental groups;
(4) a tendency to evaluate research projects in societal rather than in purely scientific terms.

In this exploratory project, we have chosen a limited number of variables which throw some light on the problem of the political orientation of scientists and on the changes in focuses of their research interests.

The sample

Years ago already eminent Dutch biologists like Kuenen (1972) showed their concern with the Dutch environmental situation. But not much is known about the attitudes and behaviour of the rank and file biologists who, though perhaps specialists in other areas, work in laboratories where ecological research projects are carried out.

There were 363 biologists involved in research and development in the Netherlands, according to the latest available figures (1969). A great majority of these biologists held a post at research laboratories (86 per cent), such as government laboratories, university and university-affiliated laboratories and other laboratories subsidized by the government.

In March 1972, a questionnaire was sent to 237 of these biologists, of which 121 biologists returned a fully answered questionnaire. After comparing the population and the sample on a number of factors (age, kind of research done, kind of laboratory and the number of biologists working there), we can say that a reasonable measure of representativeness was reached. As this is only an exploratory study it is not our aim to generalize the results.

Do we have any evidence for our main hypothesis? We shall try to answer this question in the following paragraphs.

Preliminary comments on the material collected

Almost 60 per cent of the respondents do mainly ecological research, which means that even in ecological laboratories biologists are engaged in non-ecological projects. We presume that these non-ecological projects are supporting ecological research in one way or another.

(a) *Ecological versus general projects*

In the Netherlands, most ecological projects are directed towards preservation of clean air, water and soil. As non-ecological or general projects, we have considered the projects in other areas of biology, mainly animal and plant physiology, biophysics, biochemistry, genetics, microbiology, histology, taxonomy (animal and plant), ethology and animal psychology. Ecological projects are more often the scientists' own choice than general projects, which tend to be commanded by the laboratory management.

Biologists doing ecological research give 'the advancement of science' as well as 'importance to society' as the principal motives for doing what they are doing, while biologists engaged in general projects mainly give 'the advancement of science' as their motive. In other words, concerned biologists are oriented to both the scientific community and the wider society. With regard to the existence of research groups, we found that the biologists engaged in ecological projects worked in a team more often than those who were doing general projects.

Biologists who are doing mainly ecological projects are engaged in fewer projects at a time than biologists doing mainly general projects: 40 per cent of the responding biologists were engaged in one project only, mainly ecological, 22 per cent in two projects and 38 per cent in three or more projects. Most of these projects run for five or more years, some for twenty years, and one respondent expected to be busy on a project until the year 2000!

It seems, then, that Dutch biologists are conducting much ecological research, but this does not appear to have led to a devaluation of the advancement of science as a motivating force in the choice of research topics.

(b) *Articles and lectures*

The majority of our respondents have written books, reports and articles. Biologists engaged in ecological research write articles on environmental problems more often than biologists doing general projects. The older the biologist, the more he tends to have published and the more articles he has published, the more those articles tend to be on a general biological subject. There is also a clear relationship between age and the nature of the biologist's research interests: biologists under forty are more often engaged in ecological projects than biologists of over forty years of age. We have not been able to establish a separate relationship between the holding of lectures (for lay publics) and being engaged in ecological projects. Holding lectures is probably a function of participation in environmental associations and action groups.

(c) *Associations and action groups*

A very large part of our respondents (80 per cent) are members of one or more associations for the protection of the environment besides being a member of one or several scientific and professional associations. We have divided the environmental associations into two categories: traditional and action. Traditional environmental associations are those which have been in existence for more than about ten years, are general in nature and have always had a sizeable proportion of scientists in their membership. Action groups are those which have come into existence only fairly recently, have a specific objective (like preventing the building of a road through woodlands) and a lay membership.

Biologists who are members of several environmental associations are more often members of action groups than those who are members of only a few associations. Our explanation is that many biologists are members of one or a few traditional environmental associations even before they start studying biology, and these associations probably are an important factor in the process of their occupational choice by helping to turn a diffuse 'love of nature' into a distinct academic discipline. Membership in action groups, however, may be seen not so much as a way of expressing 'love of nature', but as a means of expressing concern with a specific environmental issue, and perhaps as a reaction to an inflexible laboratory situation.

The more concerned a biologist is, the more he lets his choice of research topics—general versus ecological projects—be influenced by external sources as indicated by active participation in lay action groups and addressing lay publics as well as scientific audiences in speech and writing.

Those who are concerned are also more of the opinion that the professional training of biologists should be more ecologically oriented than it is now. The pattern of occupational choice and professional socialization in biology may have to lose its traditional character in the near future to cope with biologists' desires to take sides with environmentally concerned citizens.

(d) *Environmental concern and scientific norms*

We have operationalized three attitudinal variables: scientific norms, moral responsibility and concern about the environment, but we have not been very successful here, as we have not yet been able to validate any of the scales. We do not know for certain whether what we wanted to investigate is indicated by the items of our scales. We give the scales in the appendix. The 'scientific norms' scale especially has let us down: either those norms do not exist or our operationalization was inadequate. We shall continue to investigate this.

Discussion

We have tried to establish empirically some exogenous sources of change in research topics of biologists. Given the exceptional outburst of environmental concern in the Netherlands at the beginning of the 1970s, one would expect to find traces of this in the attitudes and behaviour of biological researchers. Or one would expect to stumble on the defence mechanisms by which the scientific community guards its position as the principal source of research interests of scientists. But this word 'or' proved to be wrong. We had to abandon one of the original hypotheses that an increase in moral concern for the well-being of society would diminish the orientation of biologists toward the scientific community. These two factors seem to be on separate dimensions and might very well strengthen each other. But we do not know, as we have not been able to validate our science scale. And, of course, we had a sizeable non-response of 50 per cent of our original sample, about which we know very little. We can only speculate that many responding biologists saw our questionnaire as yet another means to conduct action, so that this was another way of expressing concern with the environmental situation. In the same way, we can speculate that many non-responding biologists may have viewed the questionnaire as a nuisance keeping them away from their proper work: contributing to the advancement of science.

There is another factor we do not know enough about: the laboratory situation. If biologists have freedom of choice of research topics and are not restrained by the demands of their laboratory, they will tend to yield sooner to outside pressure. But the management of research laboratories—even of those working mainly on ecological projects—are often not able or willing to change the nature of the projects of their laboratories on short notice because of the structural inflexibility of research institutes, arising out of the following three factors:

(1) most of the projects are long-running, some for twenty years or more;

(2) the way projects are financed favours long-term research-programmes;

(3) the influence of existing channels of patronage in science: ecological projects tend to be sponsored mainly by government agencies, general projects by members of the scientific community like the Ph.D. advisers and laboratory heads. Adversity to government interference often means adversity to changing existing research programmes.

When environmental associations want certain topics to be investigated, their best channel would be to approach a government agency or to exert direct pressure on the biological researcher. Whether this would cause serious role strain, as implied by Cotgrove and Box (1970), is a subject for further investigation.

It is not difficult to predict that fundamental ecological research will become increasingly important as a means of coping with the damage done to the biosphere. This means that 'the future of mankind' may depend to an ever greater extent on whether biologists are willing to let ecological concern enter their decisions about their choice of research topics.

*Appendix**

The science scale

1 A researcher should know what is done with the results of his research.
2 A responsible scientist does not retain from the public any information about the adverse uses that can be made of his research outcomes.
3 A scientist should always make the result of his research generally available.
4 Science can find solutions for all societal problems.
5 Only when your publications are cited affirmatively have you really started making a career in science.
6 Scientific work need not necessarily be useful for society.
7 Science has its own laws and can ignore political barriers and the distribution of power in the world.
8 The Nobel prize is the summit of a scientist's career.
9 The results of genuine scientific work should be accepted as facts unless they can be falsified.
10 The finest thing that can happen to a scientist is getting a really original idea.
11 A researcher is not especially responsible for the welfare of humanity.
12 A scientist should publish the results of his researches only in scientific journals.
13 The only thing that matters is the advancement of science.

Moral responsibility

This was operationalized by the following items:
1 Knowing what happens to the results of your research.
2 Preventing wrong uses of the results of your research.
3 Stimulating the correct application of the results of your research.

Concernedness

To operationalize 'concern' we have tried to find in the literature an existing scale of environmental concern, or items from which such a scale could be composed. Unfortunately, we have not been able to find such items or such

* We would like to thank Ruth Shershevski, graduate student at the Hebrew University, Jerusalem (with J. Ben-David) for permission to use some of the scale items from her research project on agricultural researchers.

a scale in the literature. So the first thing was to include in our questionnaire a number of items some of which, with reasonable face validity, would probably constitute a scale, which we call the 'preliminary environmental scale'.

These items have been analysed according to Mokken's method (1970, ch. 5). Out of twenty-one items, five have been combined to form a Guttman-type scale.

Environmental scale

1 Plastic wrapping material is objectionable. 0·41
2 Biologists must be the first to develop a new ethic for the environment. 0·31
3 Biologists must warn everyone all the time that the biosystem is threatened by environmental pollution. 0·42
4 The use of the nuisance Act is an absolutely insufficient guarantee for the protection of the environment. 0·49
5 The scientist who is conscious of his responsibility must not conceal from the public any information about the wrong uses of the results of this research. 0·31

References

Barber, B. (1952), *Science and the Social Order*, Chicago: Free Press.

Barnes, S. B. and Dolby, R. G. A. (1970), 'The scientific ethos: a deviant viewpoint', *European Journal of Sociology*, 11, pp.3–25.

Ben-David, J. (1971), *The Scientist's Role in Society: A Comparative Study*, Englewood Cliffs, N.J.: Prentice-Hall.

Carson, R. (1968), *Silent Spring*, 4th ed., Harmondsworth: Penguin.

Chisholm, A. (1972), *Philosophers of the Earth: Conversations with ecologists*, London: Sidgwick and Jackson.

Commoner, B. (1968), *Science and Survival*, 5th ed., New York: Viking Press.

Commoner, B. (1971), *The Closing Circle*, New York: Knopf.

Cotgrove, S. and Box, S. (1970), *Science, Industry and Society: Studies in the sociology of science*, London: Allen & Unwin.

Crane, D. (1972), *Invisible Colleges: Diffusion of knowledge in scientific communities*, Chicago and London: University of Chicago Press.

Dieren, W. van (1972), *Handboek voor vervuild Nederland*, Amsterdam/Brussels: Elsevier.

Ehrlich, P. (1968), *The Population Bomb*, London: Ballantine Books.

Greenberg, D. S. (1969), *The Politics of American Science*, Harmondsworth: Penguin.

Hagstrom, W. O. (1965), *The Scientific Community*, New York: Basic Books.

Hancock, J. G. (1971), 'Environmental problems and the reunification of the scientific community' in Fuller, W. (ed.), *The Social Impact of Modern Biology*, London: Routledge & Kegan Paul.

Handler, P. (ed.) (1970), *Biology and the Future of Man*, New York: Oxford University Press.

Kuenen, D. J. (1972), 'De bioloog als onheilsprofeet' in Couwenberg, S. W. (ed.), *Tijdsein. Peiling en perspectief van onze tijd*, Alphen aan den Rijn: Samsom.

Kuhn, T. S. (1970), *The Structure of Scientific Revolutions*, 2nd ed., University of Chicago Press.

Lemaine, G., Lecuyer, B.-P., Gomis, A. and Barthelemy, C. (1972), *Les Voies du succès. Sur quelques facteurs de la réussite des laboratoires de recherches fondamentale en France*, Paris: CNRS.

Merton, Robert K. (1957), 'Science and democratic social structure', *Social Theory and Social Structure*, rev. ed., pp.550–61, New York: Free Press.

Mokken, R. J. (1970), *A Theory and Procedure of Scale Analysis*, ch. 5, The Hague: Mouton.

Mulkay, M. J. (1972), *The Social Process of Innovation: A study in the sociology of science*, London and Basingstoke: Macmillan.

Pelz, D. C. and Andrews, F. M. (1966), *Scientists in Organizations: Productive Climates for Research and Development*, New York: Wiley.

Ravetz, J. R. (1971), *Scientific Knowledge and its Social Problems*, Oxford: Clarendon Press.

Storer, N. W. (1966), *The Social System of Science*, New York: Holt, Rinehart and Winston.

Taylor, G. R. (1968), *The Biological Time Bomb*, London: Thames and Hudson.

Watkins, J. (1970), 'Against "normal science"' in Lakatos, I. and Musgrave, Alan E. (eds), *Criticism and the Growth of Knowledge*, Cambridge University Press.

Whitley, Richard D. (1972), 'Black boxism and the sociology of science. A discussion of major developments in the field', in Halmos, P. (ed.), *The Sociology of Science*, Sociological Review Monograph No. 18, University of Keele.

Zuckerman, Harriet and Merton, Robert K. (1972), 'Age, aging and age structure in science' in Riley, M. W., Johnson, M. and Foner, A. (eds), *A Theory of Age Stratification*, vol. III, New York: Russell Sage Foundation.

11

Aspects of the structure of a scientific discipline*

Stuart S. Blume and Ruth Sinclair

Introduction

The focus of this chapter is a single scientific discipline: chemistry. We shall seek to describe and explain certain characteristics of the social structure of the discipline, in the United Kingdom. The picture presented is essentially a 'snapshot' of the situation at a given time: which includes the distribution of research interests and characteristic differences in work organization, publication practices, and competition through the field. We use the 'sub-disciplines' within chemistry as our units of analysis.

Our discussion is principally based upon a study of research in British university chemistry departments that we carried out early in 1971 (Blume and Sinclair, 1973). Responses to a substantial questionnaire were received from 55 per cent of chemists typical (at least in terms of rank-distribution) of the overall population. The questionnaire was based upon fifty preliminary interviews. Respondents allocated themselves, on the basis of their research interests, to one or more of eight *listed* 'sub-disciplines', which corresponded to terms widely used by chemists. Any respondent unable to make an appropriate selection was invited to add a preferred term; less than 5 per cent felt obliged to do so. Previously, it had been our intention to use a much more detailed breakdown by research topics, of the sort em-

* We would like to thank the Department of Education and Science and the Council for Scientific Policy, London, for financial support of the work upon which this chapter is based.

ployed by Hagstrom (1967), but no acceptable scheme could be devised. Our sub-disciplines, accordingly, (theoretical chemistry, physical chemistry, organic chemistry, physical organic chemistry, inorganic chemistry, physical inorganic chemistry, analytical chemistry, and biochemistry) are defined by traditional usage.

The sizes of sub-disciplines

We shall first identify systematic differences in certain characteristics of the social structures of these sub-disciplines. Our objective will then be to explain these observed differences in terms of variables derived from the parameters of theoretical analysis.

Table 11.1 *Distribution of major research interests of British and American chemists*

Sub-discipline	British chemists		American chemists (Hagstrom's data)
	(percentages)		
Theoretical chemistry	6·4	(6·4)*	6·3
Physical chemistry	30·6	(35·1)	35·7
Organic chemistry	26·6	(26·6)	18·9
Physical organic chemistry	11·2	(11·2)	11·7
Inorganic chemistry	11·7	(16·1)	12·3
Physical inorganic chemistry	8·9	(—)	—
Analytical chemistry	4·0	(4·0)	3·3
Biochemistry†	3·6	(3·6)	10·2
Other	4·6	(4·6)	1·7

* Figures in brackets are derived by the division of physical inorganic chemists equally between physical and inorganic chemistry—to correspond with Hagstrom's classification in which physical inorganic chemistry could not be distinguished.
† Here we refer to biochemistry as practised in departments of chemistry. Most biochemistry research in the UK is by now carried out in departments devoted to biochemistry itself.

The distribution of the principal research interests of respondents between sub-disciplines is given in Table 11.1. Respondents were asked to indicate more than one area if their interests were, in fact, equally divided. Any subsidiary interests were to be noted separately. 10 per cent had more than one major interest and 37 per cent had at least one subsidiary interest. Table 11.1 also gives the distribution of interests of a sample of American chemists surveyed by Hagstrom earlier (Hagstrom, 1967, p.89).

The data in Table 11.1 show a reasonable agreement between the interests of British and American chemists. Although there are difficulties in comparison, both surveys used self-reporting and it may be tentatively suggested that these lists give some indication of the current relative attractiveness of the sub-disciplines as research areas. We may therefore claim that our distribution of interests is in some measure characteristic of chemistry as a whole, and not of some parochial variant.

Since many respondents had interests in more than one sub-discipline, we examined the linkages between research areas as they may be presumed to serve an integrative function for the whole discipline. To what extent do multiple interests link areas of chemistry which have long been related, whether by common technique, subject matter, or theoretical framework: conversely, to what extent do they link disparate branches of the discipline? The closer examination of a 10 per cent sample of respondents yielded forty-nine linkages (2 interests=1 linkage; 3 interests=3 linkages, etc.). Expected linkages were most common: organic/physical organic chemistry 16 per cent of all linkages; theoretical/physical chemistry 14 per cent; physical/physical inorganic chemistry 8 per cent; physical/physical organic, organic/biochemistry, physical organic/biochemistry each 6 per cent. But there were substantial minorities with interests which may be either totally unrelated or, perhaps more likely, may represent a still unorthodox application of the theories and practice of one branch of chemistry to the subject matter of another. Ben-David and Collins (1966) refer to just such a 'hybridization' process in discussing the origins of psychology as a scientific discipline. They distinguish 'role hybridization' ('fitting the methods and techniques of the old role to the materials of the new one, with the deliberate purpose of creating a new role') from 'idea hybridization' ('the combination of ideas taken from different fields into a new intellectual synthesis,' but without attempting to create a new academic role or 'giving rise to a coherent and sustained movement with a permanent tradition'). The linkages which we found (for example) between inorganic chemistry and biochemistry, and between physical and theoretical chemistry and pharmacology, suggest the importance of a 'hybridization' process within scientific disciplines. But, at this point, our non-developmental method of inquiry does not allow us to distinguish between the two variants.

Of course, scientists rarely work alone. The same sort of hybridization process may also occur through the close collaboration of chemists with different interests whether they work together in permanent research groups or on an *ad hoc* basis. In fact, a combination of expertise may be a prerequisite of progress in certain areas of research. A respondent expressed a not uncommon sentiment: 'One aspect of my work, that on metalloenzymes, requires a large group, or the collaboration of a number of groups, since a combination of expertises is necessary for a really effective attack on these complicated systems.' Thus, the nature of the problem being tackled may dictate the forms of research organization—the development of large research groups, combinations of expertise, instrumentation, etc.

Instrumentation and organization

This takes us to a description of the ways research groups working in the sub-disciplines of chemistry are, in practice, organized. We shall be concerned primarily with two issues: the size of research groups and the instrumentation employed (which may be regarded as the 'technology' of the research process). Hagstrom found these two factors related: the more sophisticated the technology, the greater the need of the principal investigator for assistants (Hagstrom, 1967, pp. 31–3). One respondent, leading the only group in the country applying a sophisticated version of mass spectrometry to ionmolecular reactions, wrote:

> I am in favour of large groups involving several established academic people and supporting groups because the level of sophistication in mass spectrometry is resulting in a steep rise in costs. The nature of the problem often cuts across disciplines: I as a physical chemist have, on several occasions sought *detailed* advice from organic chemists, physicists, electronic engineers (not technicians) and computing people. . . . As costs rise, it is uneconomic to build very similar instruments all over Britain, but equally it is unlikely, that a small group will realise the full potential of a complex piece of equipment without the stimulus of people of different backgrounds.

The use of sophisticated instruments, especially when used as exploratory tools rather than as the focus of research interest in themselves, may dictate a division of scientific labour. Swatez (1970) and Gaston (1969) have indicated how this operates in high energy physics. A similar situation exists in chemistry. Research which is essentially concerned with the elucidation of changes in a chemical system, and which employs a sophisticated instrument based upon a relatively new phenomenon, will frequently require the inclusion of a specialist in that instrumental technique within the research group.

We turn, now, to examine the utilization of various scientific instruments within the different sub-disciplines. Like Hagstrom, we found that chemists are highly dependent upon 'a wide variety of complex analytical tools', but the 'technologies' of the sub-disciplines are quite different. Virtually all organic chemists use a variety of spectrometers (infrared, ultraviolet, NMR, mass spectrometers) in the purely routine characterization of compounds. The use of such commercial instruments no longer requires the presence of an expert in instrumentation, nor is it in any sense an indication of the problem being studied. The use of these instruments is not 'problematic'. We may even say that their field of work is partly characterized by the consistent use of these instrumental techniques. (Or, as Kuhn suggests, we may say that a certain 'instrumental commitment' is an important aspect of the paradigm of the field of organic chemistry [Kuhn, 1962, p.60].) However, the same is much less true of other fields. Inorganic chemistry, for example, has no 'typical' instruments. In physical chemistry, the situation is different again; only the computer is widely used. For physical chemists, the relationship between a phenomenon and its effect upon an instrument *is* problematic. Physical chemists may be concerned to devise instruments with which they can examine a newly discovered property of a system. Research areas within physical chemistry are often characterized by their use of a particular technique: flash photolysis, laser X-ray photo-electron spectrometry, low energy electron diffraction, etc. Common technology is not, however, a unifying factor within physical chemistry as a whole. Such 'cognitive institutionalization' as it possesses must be sought elsewhere. We may note, nevertheless, that the eventual application of these new (non-routine) techniques to systems themselves defined as problematic by other sorts of chemists will follow, initially by the sort of 'hybridization' (or collaboration) process which has already been mentioned.

In short, the 'technology' of the various chemical sub-disciplines stands in a varying relationship to the kind of consensus holding the field together. In some instances, we have suggested (e.g. organic chemistry) it is an important element in that consensus. In other instances (e.g. within physical chemistry), instrumentation must be seen rather as part of the problematic of the field: the term 'technology' is perhaps not even appropriate. At the same time, instrumentation may partially dictate the social organization of research in a field.

We, therefore, turn to the size of the chemical research groups for it is a dimension of significant variation between the sub-disciplines. Table 11.2 gives the distributions in research group size. These research

Table 11.2 *Distribution of research group sizes for chemical sub-disciplines*

Sub-discipline	Numbers in group		
	25th percentile	50th percentile	75th percentile
Theoretical chemistry	2	3	6
Physical chemistry	2	4	7
Organic chemistry	2	4	8
Physical organic chemistry	2	4	6
Inorganic chemistry	2	3	5
Physical inorganic chemistry	2	3	5
Analytical chemistry	4	5	8
Biochemistry	2	5	13

groups are chiefly composed of students working for Ph.D.s (only 40 per cent of respondents had post-doctoral fellows in their groups while 20 per cent had more than one). That groups in organic chemistry are larger than those in inorganic chemistry—as well as being more numerous—must partly reflect the greater attractiveness of the field for research students. Groups in analytical chemistry are large but few in number. Whatever the causes of these differences, it is possible to speculate upon some of the consequences.

Many chemists believe that the ways large groups work typically differ from those of small ones, necessitating—for instance—the 'lieutenant' system. One physical chemist, with a subsidiary interest in biophysics, wrote:

> I can see some advantages in having groups of people with different expertise becoming 'problem-oriented' rather than

> remaining as small groups largely 'technique-oriented'. I feel that a 50:50 mixture of such types is necessary as it is only in the small 'technique-oriented' groups that novel and important instrumental application developments tend to occur. The main problem of large groups is that of who is 'boss' in both financial and academic matters, i.e. the usual problems of human relationship when progress and promotion are at stake.

As this suggests, there is a widely held view that it is in small groups that original ideas are likely to evolve while in large ones more routine problems are most effectively tackled. It is believed that the two activities are different and thus require distinctive modes of organization. One respondent summed up the general view:

> It depends on the state of development of the topic: in the initial stages a small team of post docs with good instruments and technical assistance facilities are likely to achieve more than a larger group of, say, graduate students. With most established topics, a larger team of graduate students is likely to be the more effective.

Another comment on the size of organizations comes from an eminent Fellow of the Royal Society. He discusses whether large research groups are more effective than small ones:

> It is likely to be true of certain kinds of research and of research in certain phases of development. It is likely to be true when systematic investigation is the most likely way to get results. It is not necessarily going to be true when a really new idea is needed to start progress—but when this has happened, it may be true.

It may be surmised that the closer a group approaches 'applied' problems, the more its work requires 'systematic investigation'. It is, therefore, not surprising that the two areas of chemistry in which practitioners most particularly claimed direct applicability for their work (analytical and organic chemistry) are those with the largest research groups.

Next, we attempted quantitatively to examine the relationships between the size of research groups and research performance. Are individual scientists in large research groups especially effective in some areas of chemistry, but no more than averagely effective in others? Table 11.3 shows that there are indeed substantial differences between the sub-disciplines in the association between research group size and the performance of individuals in the group. Numbers of papers published are used as a crude indicator of performance (cf. Whitley and Frost, 1971).

Table 11.3 *Research group size and performance of individuals in chemistry by sub-discipline of major research interest*

Sub-discipline	Correlation between size of research group and number of papers published over past 5 years
Theoretical chemistry	0·41
Physical chemistry	0·24
Organic chemistry	0·39
Physical organic chemistry	0·48
Inorganic chemistry	0·43
Physical inorganic chemistry	0·52
Analytical chemistry	0·56
Biochemistry	0·26

The interpretation of these figures on the basis of our hypothesis—which reflects a view common in the scientific community—is as follows. A high correlation between research group size and productivity suggests that the field is one in which results are best obtained by large teams. In other words, its typical problems are of the kind particularly susceptible to attack by large teams or they are 'routine'. Conversely, a low correlation suggests that typical problems are of a different kind, requiring more mental effort than systematic investigation.

Publication practices

Scientists in some branches of chemistry typically publish more papers than scientists in other branches. If different rates of publication are characteristic of the branches of a discipline, this is, of course, an important limitation upon the use of paper counts as a relative measure of performance at the disciplinary level. Table 11.4 gives the publication information in British chemistry with a breakdown by sub-disciplines. How may these differences be explained? An obvious answer is that it is easier to obtain publishable results in some areas (e.g. analytical chemistry) than in others (e.g. physical chemistry).

Whilst this may give some indication of the practical difficulties of successfully completing an experiment it nevertheless begs the question of what constitutes a 'publishable' result.

Table 11.4 *Mean publication rates in chemistry by sub-discipline of major research interest*

Sub-discipline	Papers published arithmetic means (of 5-year totals)	standard deviation
Theoretical chemistry	13·4	12
Physical chemistry	14·6	12
Organic chemistry	16·9	16
Physical organic chemistry	17·4	19
Inorganic chemistry	17·9	21
Physical inorganic chemistry	18·5	19
Analytical chemistry	26·4	20
Biochemistry	16·5	16

The results of chemical research are presented in a wide variety of journals. Some publish over the discipline as a whole (e.g. *Journal of the Chemical Society*, *Journal of the American Chemical Society*) and others are restricted to a sub-discipline (e.g. *Journal of Physical Chemistry*) or to a still smaller field of inquiry (e.g. *Journal of Gas Chromatography*). The reasons for the emergence of these more specialist journals have been discussed by Hagstrom (1965, ch.IV). According to his study, a major stimulus for specialist journals was a belief within a section of a disciplinary community that their own particular goals, problems, theories, were inadequately regarded. The 'deviant' specialists complain that they receive insufficient journal space. They also may reject the criteria which determine editorial evaluations (Hagstrom, 1965, p.202). In the new journal evaluations will be made according to criteria defined by and for the new specialism and are naturally different from those common in the discipline as a whole. The criteria which govern the acceptance of a paper by a discipline-wide journal differ in scope from those pertinent to its acceptance by a more specialist journal.

Furthermore, Hagstrom (1965, p.210) has suggested that new journals may find it difficult to impose standards equal to those of the older journals. If we compare the specialist journals which have evolved within and around such a well-established discipline as chemistry, we may anticipate certain other differences. For example, we may hypothesize that the more complete the articulation of a paradigm, the greater the agreement over the critical problems of the field, whether this agreement derives from basic theory or from some extra-theoretical aspect of the paradigm. The greater the extent of this agreement, the smaller the desire (or the opportunity) to publish on a wide range of topics. Perhaps, here, we have the explanation of differences in typical productivity between the sub-disciplines. Physical chemistry and analytical chemistry both possess their own periodicals (as well, of course, as access to the more general journals). Perhaps, the greater average productivity of analytical chemists may be explained by a lower degree of consensus hence by less rigorous criteria of evaluation in that field.

The sub-disciplines are not wholly autonomous. Although the cohesive forces within each may be stronger than the forces holding them together, they are nevertheless united both cognitively and socially (Whitley, ch.4, above) in some measure. The term 'chemistry' is a meaningful one. We have already suggested that its sub-disciplines may differ both in the 'routineness' of their activities and in the rigour of the evaluation process applied by their own journals. But is it possible to say anything about their relative contributions to the development of the discipline as a whole? The evidence which we shall now present relates to the citations typically received by papers in the different branches of chemistry.

In 1963–4 some 6,000 papers appeared in the twenty-seven journals published in the UK which we consider as chemistry, biochemistry, pharmacology, and industrial/applied chemistry. In 1965, according to the *Science Citation Index*, these papers received 20,300 citations—an average of 3·4 each (Martyn and Gilchrist, 1968). But there were significant differences between the sub-disciplines. The data in Table 11.5 give the average number of citations in 1965 to papers published in 1963–4 in journals clearly within one of the sub-disciplines. Journals covering all chemistry appear under 'general'.[1] It should, also, be noted that the majority of articles published in biochemical and pharmacological journals are neither read by nor used by chemists.

The full explanation of these differences may contain elements of

[1] Excluded from the tabulation is the general *Quarterly Review*, whose fifteen review articles received 243 citations, an average of 16·2 per article. The importance of this review literature is apparent (cf. Price, 1970).

two partial explanations: one having significance principally for the empirical study of the social organization of science, the other relevant to our understanding of the cognitive structure of science. In the first place, since the sub-disciplines of chemistry differ in size, to the extent that they are cognitively self-contained, the potential number of citations any paper may receive in succeeding papers will differ. The larger the sub-discipline, the larger the possible number of citations.

Table 11.5 *Citation rates for 'typical' papers in the sub-disciplines of chemistry**

Field of publication	no. titles	no. papers published (1963–4)	no. citations (1965)	ratio citations/ papers	range
General chemistry	3	1515	5318	3·5	3·0–5·3
Theoretical/physical chemistry	8	914	3439	3·8	1·5–6·6
Organic/physical organic chemistry	2	1034	3638	3·5	—
Inorganic/physical inorganic chemistry	1	349	597	1·7	—
Analytical chemistry	1	145	302	2·1	—
Biochemistry/ pharmacology	6	1276	5577	4·4	1·6–5·6
Applied/industrial chemistry	6	761	1451	1·9	1.1–5.2

*Data adapted from Martyn and Gilchrist. (1968)

On the other hand, the sub-disciplines of chemistry are not wholly autonomous—there is plenty of evidence for that and many chemists work in more than one sub-discipline. Papers may have implications outside the specialist area to which they are principally addressed. For example, the first papers describing the preparation of the fluoride of a so-called 'inert gas' had implications for the understanding of the nature of valency (theoretical chemistry), offered a new type of substance for study by physical chemists, as well as a preparative technique with which inorganic chemists could seek to make homologous compounds. Thus, a paper of high quality may serve the development of chemistry as a whole or, in Kuhnian language, may serve to further articulate that more general paradigm which defines the limits of chemistry as a discipline. It may, therefore, receive many more citations than the size of its immediate audience suggests.

We are not in a position to distinguish the relative importance of these two factors, nor will the data of Table 11.5 sustain any weighty theoretical formulation. It may be significant, though, that the most application-oriented subdivisions of the field (applied and analytical chemistry) in which authors are especially prolific prove to have rather low citation rates. At the same time, physical chemistry papers are very highly cited—more highly, indeed, than the discipline-wide journals in which chemists might have been expected to publish their best papers. The data are at least suggestive of some relationship between three important variables: the citation rate typical of papers in a sub-discipline, the rigorousness of the criteria with which papers in the sub-discipline are evaluated, and the contribution of that sub-discipline to the advancement of chemistry as a whole.

Competition among scientists and its consequences

Having identified differences between the sub-disciplines, we wish to discuss behavioural variations of the chemists. James Watson's book *The Double Helix* (Watson, 1968) and its dramatic evidence emphasized the spirit of competitiveness which pervades much of science. In this section, accordingly, we shall discuss the incidence of this important factor, competitiveness, in the various sub-disciplines of chemistry and point to some of its consequences.

The phenomenon we wish to discuss is competition between scientists rather than between paradigms or 'research programmes' (Lakatos, 1970). In the essentially Mertonian scheme, competition derives from the premium which is placed upon obtaining the first solution to a scientific problem. Recognition in science normally accrues almost solely to the first presentation of a solution. Merton has, of course, discussed, the acrimonious disputes which have sometimes followed the simultaneous discovery of a phenomenon (1961). In Hagstrom's words (1965, p.70): 'Competition results when two or more scientists or groups of scientists seek the same scarce reward—priority of discovery and the recognition awarded for it—when only one of them can obtain it.' Competition of this kind is seen as functional for science; it demonstrates the commitment of scientists to the reward (recognition) over which the scientific community itself has control. In particular, it may serve to ensure an optimum allocation of scientific effort among research problems. Large numbers of scienists, perhaps especially of good scientists, will be attracted to research areas promising particularly significant results (Hagstrom, 1965, p.81; 1967). The amount of recognition available is a function of the importance of the discovery or problem solution for science. Of course, many scientists may prefer to avoid the intense competition of these central areas. Like the businessman, the scientist has to choose between high risk-high profit and low risk-low profit strategies.

Finally, areas of science may differ among themselves in their competitiveness. Hagstrom has hypothesized that the prevalence of competition in a field of science will be greater (1) the more complete the agreement between those working in it on the relative importance of scientific problems, and (2) the larger the number of specialists possessing the skills and equipment necessary to tackle any problem in the field (Hagstrom, 1965, p.73). Kuhn's notion of the paradigm (Kuhn, 1962, 1970; see also Lakatos and Musgrave, 1970) allows some elaboration of the first of Hagstrom's criteria. We may say that the stronger the paradigm, the more completely articulated, the greater the agreement on the significance of problems will be. Similarly, the weaker the paradigm the more scientists may be expected to turn outside the field either for problems to solve or for explanatory hypotheses. Under such conditions the mere fact that scientists share a field of research will be no guarantee of agreement. Indeed agreement over what constitutes a solution, or a discovery, may not be forthcoming.[1] This hypothesis may be invoked in discussing differences between the sub-disciplines of chemistry. We have, indeed, already introduced a notion corresponding to the first of Hagstrom's factors in discussing the publication practices of scientists.

A number of possible responses to competition, or to the fear of competition have been distinguished. High competition is likely to lead to the 'anticipation' of the work of many of those involved: one scientist is likely to emerge as clear winner and scoop the pool while

[1] There has, for example, been a dispute between Berkeley and Dubna physicists over the discovery of element 105. The American group's leader has written (Ghiorso, 1971): 'I would like to raise the question of what constitutes the discovery of a new element. It seems to me that the discoverer is the one who first proves that he has indeed found a new element. Our published work demonstrates beyond question that we have identified the isotope $^{250}105$ by linking it genetically to its well-known laurentium daughter ^{256}Lr. . . . On the other hand the Dubna discovery of a 2-second spontaneous-fission emitter is still open to question as to the identity of the atomic number involved.'

the others find that years of effort reap no reward. Some scientists avoid this possibility. As Gaston writes (1971), on the basis of his study of British high energy physicists, 'what is certain is that a large proportion of the scientists who have never been anticipated have withdrawn from, or consciously chosen not to engage in, the more competitive areas'. Similarly, Ben-David and Collins have suggested that a major stimulus to the development of new fields of research is provided by the migration of scientists from traditional fields, in which 'the chances of success . . . are poor' to related fields 'in which the conditions of competition are better' (Ben-David and Collins, 1966). Avoidance, or migration, is one possible response to competition. A second strategy distinguished by Gaston, although said to be relatively rare, is fraud or theft of the results of others (Gaston, 1969, 1971). A third is secretiveness. Scientists may avoid freely discussing their current work for fear that others may be inspired to turn to the problem, and perhaps through having access to greater resources, solve it first. A fourth strategy which is sometimes adopted is the early publication of partial results in the attempt to 'stake a claim'. As Hagstrom (1967) has shown, disciplines differ in the extent to which the publication of abstracts in this way is permitted or encouraged: in physics 'rapid publication journals' such as *Physical Review Letters* are used specifically for this purpose. A fifth and final means of staving off the possibly disastrous effects of competition is collaboration. Scientists who discover that they are working on aspects of the same problem may decide—especially if they know and trust one another—to divide the problem between them, in order to ensure each a share of the credit, or otherwise collaborate.

Following Hagstrom (1967), experience of competition was defined in our study by the number of times in his career an individual had been anticipated. We also introduced his concern about competition (being anticipated), defined by his answer to the question: 'How concerned are you that you might be anticipated in your current research?' Among the possible consequences of competition (or of concern about competition) secretiveness derived from responses to the question: 'Would you feel quite safe in discussing your current research with others doing similar work in other institutions or do you think it necessary to conceal the details of your work from some of them until you are ready to publish?' A subsequent question asked respondents to indicate what percentage of their research results were first published as preliminary accounts or letters to the editor.

Table 11.6 gives the intensity of competition in each of the sub-disciplines (biochemistry excluded). Defined in this way, competition is much less intense in physical or theoretical chemistry than in analytical chemistry. Theoretical chemistry and analytical chemistry

Table 11.6 *Relative intensity of competition in sub-disciplines of chemistry*

Sub-discipline	Percentage of respondents in the sub-discipline who have been:	
	anticipated at least once in their career	anticipated at least 3 times in their career
Theoretical chemistry	53	12
Physical chemistry	61	12
Organic chemistry	79	20
Physical organic chemistry	68	16
Inorganic chemistry	75	19
Physical inorganic chemistry	78	18
Analytical chemistry	77	19

are of roughly equal size. Physical chemistry is much larger. Moreover, in order to explain the greater productivity of analytical chemists, we have already hypothesized that the degree of consensus over critical problems in analytical chemistry must be lower than in physical or theoretical chemistry. We are caught in a dilemma. With a certain dexterity of interpretation, the explanatory variable which we have invoked (degree of consensus over critical problems) may be used to explain either relative publication rates or relative intensity of competition. It cannot explain both. We have explored this dilemma in detail since it is to our minds symptomatic of the difficulty in using cognitive variables in interactional studies at a macro level. As to its resolution we think it is our suggestion of greater consensus in physical chemistry which was in error—as is witnessed by the large number of periodicals published in that field. But, it is interesting to note that the consensus in analytical chemistry must derive less from any high level theory than from agreement over very specific extra-theoretical priorities. Characteristic differences in productivity, then, remain to be explained.

We now present our remaining data on concern about competition and its consequences as they are reflected in the behaviour of typical chemists in each field. In doing so, we are well aware that they are not readily explained within the framework we are trying to sketch. It is reasonable to assume that individuals will be affected in their subjective attitudes by the likely effects of having been anticipated; for example, by the extent that having been anticipated affects their careers. The extent to which anticipation precludes future publication may also be relevant and this Hagstrom has shown (1967, p.7) varies between the disciplines of science. Perhaps it also varies between the sub-disciplines of chemistry.

Table 11.7 *Secretiveness and pre-publication in chemistry by sub-discipline*

	Secretiveness			*Pre-publication*		
	% who would discuss work with			% work published in preliminary form		
	few/none	most	all	50	11–50	10
Theoretical chemistry	14	51	34	6	28	66
Physical chemistry	9	58	33	6	29	65
Organic chemistry	15	63	22	17	52	25
Physical organic chemistry	15	55	30	8	60	32
Inorganic chemistry	17	64	19	13	53	34
Physical inorganic chemistry	5	70	25	2	39	59
Analytical chemistry	14	79	7	4	19	77

Table 11.7 indicates the extent to which certain responses to competition, or fear of competition, are found in the various sub-disciplines. On both criteria, theoretical chemistry, physical chemistry, and physical inorganic chemistry appear least affected by the 'pathological' symptoms of secretiveness and 'claim-staking'. Organic chemistry and inorganic chemistry are most affected, although the effect is a slight one. It seems likely that the more readily the solution

to a 'typical problem' of a field may be replicated on the basis of partial information the greater will be the protectionism. Thus, it seems possible that such features of scientific fields as those illustrated in Table 11.7 may be a partial consequence of the problems, techniques and standards typifying the fields.

Conclusions

In this chapter, we have described certain features of work organization, publication patterns (and the relations between these two), and competition, as they occur in British chemistry. We have tried to point out systematic variations in these features within chemistry, taking as our unit of analysis the 'sub-disciplines' of the field. Our description has been an empirical one in which data have largely derived from a postal survey of chemists in British universities. A number of explanatory variables have been introduced in the attempt to account for the systematic differences found. In selecting these variables, it has been our intention to seek some tentative links between the 'neo-Kuhnian' analysis of the structure of science and the kind of data obtainable from a large-scale study of social aspects of science. Using the concepts of Kuhnian and neo-Kuhnian analysis in the empirical attempt to relate cognitive and social structures is extremely difficult, not least because of the ambiguity of many of the terms. Moreover, the attempt to operationalize those terms, as in Whitley's chapter in this volume, largely revolves around rather small segments of disciplines ('research areas,' 'specialties') defined, as it were, with Kuhnian theory in mind, and small enough to be susceptible to sociometric analysis. Our own work has shown that whole disciplines cannot be studied in that way—individuals cannot be assigned unambiguously to one of a list of mutually exclusive specialties. We have, therefore, introduced some cognitive variables less well-rooted in theory, but which we feel are at once utilizable at the disciplinary level and relatable to those employed in more detailed analysis of scientific research areas.

References

Ben-David, J. and Collins, R. (1966), 'Social factors in the origins of a new science: the case of psychology', *American Sociological Review*, 31, pp.451–65.

Blume, S. S. and Sinclair, Ruth (1973), *Research Environment and Performance in British University Chemistry*, London: Department of Education and Science (Science Policy Study, No. 6).

Gaston, J. C. (1969), 'Big science in Britain: a sociological study of the high energy physics community', Ph.D. dissertation, New Haven: Yale University.

Gaston, J. C. (1971), 'Secretiveness and competition for priority of discovery in physics', *Minerva*, IX (4), pp.472–92.

Ghiorso, A. (1971), 'Disputed discovery of element 105', *Science*, 171, p.127.

Hagstrom, W. O. (1965), *The Scientific Community*, New York: Basic Books.

Hagstrom, W. O. (1967), 'Competition and teamwork in science' (mimeographed), Madison: Department of Sociology, University of Wisconsin.

Kuhn, T. S. (1962), *The Structure of Scientific Revolutions*, Chicago University Press.

Kuhn, T. S. (1970), *ibid.*, rev. ed. with additions.

Lakatos, I. (1970), 'Falsification and the methodology of scientific research programmes' in Lakatos, I. and Musgrave, Alan E. (eds), *Criticism and the Growth of Knowledge*, Cambridge University Press.

Martyn, J. and Gilchrist, A. (1968), *An Evaluation of British Scientific Journals*, London: Aslib.

Merton, Robert K. (1961), 'Singletons and multiples in scientific discovery: a chapter in the sociology of science', *Proceedings of the American Philosophical Society*, 105, pp.47–86.

Price, D. J. (1970), 'Citation measures of hard science, soft science, technology, and non science' in Nelson, C. E. and Pollock, D. K. (eds), *Communication Among Scientists and Engineers*, Lexington, Mass.: Heath.

Swatez, G. (1970), 'The social organization of a university laboratory', *Minerva*, VIII, pp.36–58.

Watson, J. D. (1968), *The Double Helix*, London: Weidenfeld & Nicolson.

Whitley, Richard D. and Frost, P. A. (1971), 'The measurement of performance in research', *Human Relations*, 24, pp.161–78.

12

Science is a social-psychological activity —even for science students

Dorothy S. Zinberg

As science, and the technology it generates, continue to evolve into increasingly powerful, if controversial, forces in society, the interest of social scientists in their institutional growth and development intensifies apace. A vital component of the growth of science is the production of scientists. Unlike literature where an untutored, creative talent can burst forth with original insights and achievements, science—increasingly complex—boasts few amateurs or later-life recruits. Scientists usually make the decision to study science between the ages of thirteen and fifteen (a significant number do so even earlier), and although almost 50 per cent move away from the decision by the time they graduate from university, relatively few non scientists move into science during the undergraduate years. Therefore, the outer limits of the pool of potential scientists are relatively fixed during this period. By understanding the training and socialization of neophyte scientists —in this study, undergraduates in an English university department of chemistry—it is possible to identify attitudes, values and expectations that separate aspiring scientists from senior scientists as well as those that they share and that appear to be maintained over time.

As John Ziman, a theoretical physicist, has written (1968), 'scientific work is only meaningful in the social context of the scientific community.' A university science department and the students it trains provide the social context that will to a large extent determine the health and vigour of the larger scientific community. Here I shall focus on one aspect of the community: the current generation of science undergraduates, the majority of whom, in the chemistry department studied, do not resemble the stereotypic image of the science student or scientist.

A three-year longitudinal study[1] of chemistry undergraduates at an English university demonstrated that rather than pronounced inclinations toward social isolation, seriousness and introspection—qualities most often associated with scientists—the students emphasized social process—making friends, working with people and learning about themselves and each other. The majority believed it was 'essential' that a university education help one 'learn to get along with people' rather than 'get a job'.

The discrepancy between the anticipated findings of the study—science students would be more interested in things, ideas and vocations than in people—and the actual findings—they preferred people and people-related activities over all other factors—raised the interesting question about the possibility of a newly emerging social-psychological profile of undergraduate science students. Because a significant shift in attitudes, values and behaviour will have important repercussions for the creative potential of the scientific community, it is useful to explore the general direction of this shift so that the scientific community can develop curricula and learning environments that will respond to the educational needs of an increasingly heterogeneous group now somewhat too stringently defined as science undergraduates.

Background

In 1968 I began an informal participant-observation study of a chemistry department in an English university in part as a response to the Dainton Report (1968) which warned of an impending shortage of scientists and science teachers and which, in addition, pointed to the lack of data about the personality of science students. Would knowledge of student personality be useful in attracting more students to science? Would understanding of the factors that led to students dropping out of the course be advantageous in decreasing the attrition from science?

Although economists and science policy planners might have been concerned with future science-manpower shortages, those who had chosen science as a career looked upon the imbalance between supply

[1] The results of the first part of the study have been published in *Science Studies*, where a detailed account of the methodology is included (Zinberg, 1971).

and demand as auspicious for their future careers. A study published in 1967 caught the optimistic mood then shared by students and staff who had chosen science as a profession (Hutchings, 1967, p.11):

> Many will prefer to stay on in university appointments (rather than enter industry) and with current expansion there will be plenty of such opportunities available. Indeed one of the main reasons for expanding postgraduate facilities advocated by Robbins was to increase the source of university teachers. While this should pay off in the long run, it seems inevitable that, for some time to come, industry and the nation's schools will find it increasingly difficult to recruit the scientists they so badly need.

Although there was little evidence of enthusiasm among the entering class for chemistry as a subject, students did believe that on graduation they would have the choice among several good jobs or postgraduate courses.

I, too, shared the concern of science educators with the reports of impending shortages of scientists in teaching and industry. I also shared the incoming students' expectations of an open job market and career opportunities limited only by one's ambition and drive. Both assumptions were rudely shattered within weeks of beginning the study as reports of cutbacks in postgraduate openings and reduced hiring by industry began to filter down to the new arrivals in the department—the students and myself. Science education would be examined by students in a new light—a dimmer one.

During the first year of the study I joined the entering class of chemistry students at the university. Armed with a tape recorder and notebook, I listened, watched, talked, recorded and gossiped with a majority of students, staff and administrators. I interviewed a random selection of students and staff at intervals throughout the academic year. From the interview data, I constructed questionnaires which were administered to the entire first-year class at the beginning of the first term and at the end of the second.

Among the goals I had were, first, to delineate the attitudes of the students to the courses they were taking and, second, to compare their perceptions of what they were learning with what the lecturers thought they were teaching. This goal was deflected somewhat by the students' responses and informal conversations. They wanted to talk about boredom and their classmates; the contents of the lecture were of lesser concern. I put this down, in part, to the situation. They were new students; for the majority it was their first year away from home, and in many cases, their expectations of university were not matched by the reality of their new experiences—the emphasis might change in time.

In the spring of 1971, shortly before the students took their final examinations, I returned to the univerity for three weeks. I re-interviewed some students, chatted informally with others, and visited residence halls in the evenings and the student union during the day. In addition I interviewed several of the staff who during my absence had been promoted or had taken on new assignments (such as departmental tutor) as well as those whose interests in departmental politics or academic reform (not necessarily separable) had brought them closer to decision-making councils. I sent a questionnaire based on these interviews and observations to the entire third-year class. The discussion which follows is based on interview/questionnaire responses from approximately 85 per cent of the class of fifty-eight students.

Orientation

As a former chemist I was sensitive to the criticism levelled against social science explorations in the physical sciences that too much attention is given to social process while too little is directed to qualitative assessments of what is taught and learned. I believed that my knowledge of chemistry, even though diminished during the previous decade, would contribute to my learning more about the students' response to their work and the staff's goals for their students. Specifically, I hoped to achieve a balance between affective and cognitive data. Therefore, the interviews and the questionnaires were designed to elicit both types of information and to achieve a balanced picture of formal learning on the one hand and informal experiences on the other. Process would not overshadow substance. However, the interviews and the questionnaire responses did not reflect the intended balance and gave support to the notion that it is not social scientists alone who are more interested in social processes than in subject content; they have been joined by science undergraduates.

The stereotype

The social stereotype of the scientist is characterized by a lack of interest in people. Whether the observations stem from studies by social scientists of science students and senior scientists or from surveys of non scientists and their image of the scientist, the results usually stress the low interest the scientist has in interpersonal relations.

'The major differences between those who go into natural science and those who do not is the extent and nature of the interest or lack of it in persons.' This statement by Roe (1959, p.110) has been one of many similar guidelines for delineating the social psychology of scientists, whether as students or as practitioners. Other, more recent research has shown that their thinking patterns tend to be convergent rather than divergent; they are likely to be introverts rather than extroverts; they develop later socially, spend more time alone, and tend to be rather conventional and conformist.

Nothing I have observed or recorded suggests that these stereotypic traits are not present among some members of the group. However, the rapid growth in the numbers of students studying chemistry (the department has doubled in size since the Second World War) has led to the decrease in the percentage who resemble the typical scientist. More than 60 per cent of the chemistry students in this study are not oriented toward a career in science but rather toward a university degree that will lead to a job or career which might or might not be in science. The data suggest that although the social-psychological differences between groups of science and non science students undoubtedly persist, the within-science group variations now range over a broad spectrum of personality and sociocultural variables making the within-group variations as marked as those between groups. In addition, there is substantial pressure for science students to become more like their arts peers rather than their traditional science peers. To be judged a 'real scientist' by one's fellow students is to be found wanting.

Questions and answers

My original study was designed to understand the processes by which a student begins to evolve into a scientist. It was not intended to explore the possibility of a new configuration of personality variables among undergraduate chemistry students but rather to explore student relationships with each other, the staff, administrators and the curriculum. However, the repetition and the intensity of people-oriented answers necessitated a re-examination of the science-student stereotype.

During their first undergraduate year the majority of students expressed anxiety about becoming 'narrow' as a result of studying chemistry. By the third year, these fears were only partially realised. Although for many the educational experience had indeed proved 'narrow', the non academic experience had made the three years worthwhile.

In both the interviews and questionnaires, whenever a question was open to different interpretations and several answers were possible, students most often chose to report non academic experiences rather than strictly academic ones. For example, a specific question, 'What topics were covered in the biochemistry lectures?' would elicit a substantive answer. But a more vaguely worded question, 'How would you evaluate the biochemistry lectures?' brought forth answers about the lecturer's personality and skill as a speaker or, often, the enthusiastic response that biochemistry had been the most interesting of all during the three years because 'it related to life'.[1] Given an open-ended question, the majority of students stressed relationships with friends and life at university over any topic or course. This is not quite the stereotypic picture of the science student.

The question:

'How have the following contributed to the ways in which you are prepared for the future?'

(a) experience in the scientific method
(b) the substance of the chemistry course
(c) the non academic experiences you have had

elicited almost unanimous agreement. The experiences in scientific method had contributed 'substantially' to the students' ability to think 'logically'. Oddly enough, the one student who remarked he had had little 'individual experience' of the 'scientific method' during his three years received a First and plans a research career in chemistry.

Just the reverse was indicated by answers to the question about the contribution of the course: '95 per cent will never be needed again for most people', 'negligibly', or 'very little, indeed'. Only two students thought the substance of the course contributed to their futures, while one lone student exclaimed jubilantly, 'it gave me a degree!'[2]

However, the last question, about the contribution of 'non academic experience', prompted enthusiasm from even the most

[1] Biochemistry was chosen as the 'most favourite topic' followed closely by computer techniques ('you can tell if you're right or wrong').

[2] The most consistent criticism of the course was that it was overburdened by useless empiricism: 'too many facts regardless of what I do in my career'.

restrained interviewees and monosyllabic questionnaire respondents:

'Vital!'

'Yes, that's what life is all about.'

'The important part of growing up.'

'My personality changed.'

'Made me a more interesting person, hopefully.'

'For me, life revolves around friends. There is nothing else.'

'Life is a series of interactions with others. These were my most important experiences.'

In somewhat the same vein, I rephrased the question later in both the interviews and the questionnaires: 'What has been the most interesting aspect of your life here?' With one exception the answers were entirely social:

'My friends.'

'Stimulation of company of fellow students.'

'Life in Hall and at our house.'

The one deviant wrote 'chemistry', and after that, 'meeting people'. However, the tapes of the interview revealed that she had already enrolled in a secretarial course and had no intention of doing anything related to science in the future: 'The secretarial course is a means of getting abroad. I'm also getting interested in politics. Anything's better than doing chemistry.'[1] Another question: 'What do you talk about with your friends?' brought only five responses that mentioned science or chemistry, and two of those were qualified, 'before exams only'. Again, friends ranked first. As one young man noted, 'If X is elsewhere, we talk about X'. The other topics listed were: 'life in general', 'sex', 'sports', and 'religion'. One young man volunteered: 'If you want to keep your friends, you stay off chemistry.'

Another open-ended question, 'Have your attitudes changed in the past three years?' could have been interpreted within the context of the interview and questionnaire, as meaning either attitudes toward chemistry, science, a career in science or attitudes toward non scientific topics. Eighty-five per cent believed their attitudes had changed, but only one person mentioned anything vaguely connected with science and that was negative.[2] One student wrote: 'I am no

[1] Here is an example of the advantages of a multi-method study. Without additional knowledge of the young woman, the questionnaire response(s) would have been misleading.

[2] She believed that she now felt 'the public had been "had" because of the addition of toxic chemicals to food products' implying that her view of the chemical industry had become more cynical.

longer interested in straight A's—now people are central to my happiness.' In the same vein, *everyone* else who answered mentioned: 'personal relationships' and 'life in general'; and they agreed that they had become: 'more broadminded, less intolerant', 'more liberal', 'more sceptical', 'more understanding of other people', 'more self-confident', 'more able to communicate with other people'. A few believed they had moved toward 'cynicism and despair' and 'from bland acceptance to strong anticlericalism' in their religious beliefs. One student summed up his changed attitudes by saying: 'I have become more sensitive to other people's feelings. Just growing up. It has nothing to do with University.' Another wrote, 'My private life has been much more interesting and meant more to me than my academic life.'

All this does not imply that students never thought about their professors or their courses. That everyone who sat for the examinations received a degree demonstrates that some sort of academic work was always in progress. And when students were asked specifically about their professors, their replies were often long and detailed. Many professors received high praise:

'Some with their enthusiasm, others with scholarship, and others with their ability to put their subject across [had been most impressive].'

'Dynamic scientist, gently spoken, concise, a sympathetic face.'

'Every inch a superior person.'

Yet the majority of the favourable descriptions had little bearing on whether or not the professor was a good scientist or teacher; rather they highlighted a staff member's capacity for establishing social contact:

'Sound understanding of our difficulties, prepared to help and unique ability of appearing to enjoy lecturing.'

'Always willing to help, listen and advise.'

'Unselfish, friendly, humorous, compassionate. Will help anybody.'

'Interested in students.'

When the students were asked to rate the importance of different variables in a university education or a future job, more thought it 'essential' that a university education help one 'learn to get along with people' than that it help one 'get a job'. More than 50 per cent of the class considered that 'working with people' would be essential in a job, while less than 25 per cent felt it was 'essential' to make use of one's skills. More than 40 per cent expressed a major preference for 'human interest' in their work, while less than 16 per cent cared about 'making a contribution to science'.[1]

It was clear during the early stages of the study that the interpersonal issues were overriding. Fearing that the results might stem from an imbalance in the questions, I asked several distinguished senior scientists what they remembered of their undergraduate experiences and how I might elicit from the students some sense of the balance of social and academic thinking. They suggested I ask the students to write or tell me what they were likely to think about when they were by themselves. They all assumed that because many of their own ideas had stemmed from quiet, solitary moments spent musing, many students would be likely to report similar experiences. Although the lectures themselves might not have caused much comment, they felt that the pleasures of scientific thinking would have found a form of expression. After all, Kekulé's vision of the carbon ring had come in a dream, and more recently, James Watson's book, *The Double Helix*, had shown how he spent his idle moments on trains and elsewhere musing over conversations with Francis Crick trying to figure out what the shape of the DNA molecule might be.

The students, however, reported no musing or idle daydreaming about science. The very nature of such a question when asked of college students today is likely to evoke a ribald response which they feel quite free to enunciate. One has to ask if in today's relentlessly frank society, a student would have to report he was thinking about sex even when what he actually was thinking about was molecular structure. And, of course, one has to ask what a now distinguished scientist might have answered when he, too, was an adolescent.

Of the male students who answered the question (a few omitted it and one wrote, 'I doubly refuse!'), more than half gave responses specifically about sex, whether expressed directly (there were no rude answers, although the temptation must have been strong for several); romantically—'marrying the ideal girl'; or matter-of-factly—'girls are another well-placed topic of thought'. The remainder of the answers touched on 'anything', 'travelling', 'money', and one young man's solitary preoccupation, 'dying'.

The single reference to science by a male student came from some-

[1] In addition to questions designed specifically for the study, many questions developed by other investigators were employed in order to obtain comparative data with larger or more diversified samples. Among the authors drawn upon were D. Armstrong, B. Barnes, M. Rosenberg, D. Hutchings and R. and R. Rapoport.

one who said he spent his time daydreaming about 'the future . . . particularly science fiction'. Only one young woman mentioned science, but listed her boyfriend first. Again, because of the multi-methods relied on for gathering the data, I could assess by listening to the tapes and rereading our correspondence that her 'science' entry stemmed more from a sense of what should appear on a questionnaire about science education than from a true rank order. Perhaps science was indeed second in her thoughts, but one would have to know the relative times involved. As a group, the young women were less detailed about reporting daydreams. Only 25 per cent mentioned a boyfriend, while the majority emphasized time ('past, present or future') or were non specific—'anything or everything'.

Discussion

What then can be deduced from the personal and interpersonal observations and insights provided by the students? First, it must be stated that they are partial insights—not the total social-psychological picture of even this small group of science students. Each research method casts a mould that predetermines to some extent the form and emphasis of the data and observations it can yield. Given the same department and the general topic of the socialization of the science undergraduate, what was observed, measured, or in any way assayed would depend on the researcher's orientation to and underlying assumptions about the subject.

The statements in this essay provide one piece to be wedged into an already complicated jigsaw, assembled by social and natural scientists. What we as yet do not know, is how critical or how large this newly emerging emphasis on interpersonal relations will prove to be. However, it is possible to view the findings against the background of contemporary social change, some of which can be documented, other aspects of which are more elusive.

What can be documented is the meteoric rise in the numbers of students attending institutions of higher education since the Second World War. Barring a cutback in government funding of students the British expect to double their 1972 university and polytechnic population by 1980. This growth, which is viewed as a measure of progress, has brought about many changes. In the department of chemistry discussed here, the undergraduate population has doubled since 1948. Only a shortage of qualified students has kept it from trebling, but if programme and curriculum revisions currently under way are successful, more students will be eligible for admission. The reasons the department is eager to grow have been discussed elsewhere (Zinberg, 1971); for our purposes here, it is necessary to think only about what the impact of the unprecedented growth has been in producing the shift away from the stereotype.

When the department was smaller, the students were a more homogeneous group: homogeneous in aspirations and goals. The study of chemistry was preparation for a vocation or career in industry or at a university. More recently, chemistry (or any science) has been for many a way to obtain a university degree rather than specific training for either academic careers or industrial work. In the 'revolution of rising expectations' the goal has become the degree, not the substance or the training provided by the course. Although some students in the chemistry department are very much in the tradition of the students of past decades in personality and in occupational intentions, these are no longer the core group, but rather a distinct minority.

The other students, who have come to resemble their arts and social science peers in their emphasis on the importance of interpersonal relationships, form the critical mass. They create the social environment in which science is taught. Not only are they scornful of the 'typical' scientist ('narrow', 'a bore'), but they exert pressure on those who resemble the 'typical' scientist to broaden their horizons and extracurricular interests. Even the most inhibited and reticent of the chemistry students expect the university to provide a total experience, not simply a vocational passport. Changing mores, child-rearing practices, and educational innovations have raised students' desires to experience feelings and to know themselves. Where the idea that the university should provide the wherewithal for self-discovery and emotional growth would be and is anathema to many of the professionals who teach them, to the students it is viewed as a primary function and obligation.

As the notion of the isolated, unfeeling scientist is disparaged by science students, so the concept of a value-free, autonomous science responsible only to itself is increasingly challenged by the larger society. They are not unrelated processes.

When individuals change, the community changes. It is too early to tell what impact the younger generation of 'social' natural scientists will have on the scientific community and its creative output. But as Ziman (1968) has pointed out in his discussion of the social character of the scientific life, 'The health of the scientific community depends upon the details of the social transactions of its members.' The observations from this group of science students provide some of the vital signs and symptoms of the 'social transactions'. How they are diagnosed and prescribed for by those responsible for the education of future generations of students will determine to a large extent the vigour and well-being of that community.

References

Dainton (1968), *The Flow into Employment of Scientists, Engineers and Technologists* (Dainton Report), London: HMSO, Cmnd 3760.

Hutchings, D. (1967), *The Science Undergraduate: A Study of Science Students at Five English Universities*, Dept. of Education, University of Oxford.

Roe, Anne (1959), *Science Begins at Home*, New York: Thomas Alva Edison Foundation.

Ziman, J. (1968), *Public Knowledge: The Social Dimension of Science*, Cambridge University Press.

Zinberg, Dorothy (1971), 'The widening gap: attitudes of first year students and staff towards chemistry, science, careers and commitment', *Science Studies*, 1, p.287.

13

The science of science as a new research field and its functions in prediction

János Farkas

The relation between the science of science and the sociology of science and their role in social planning

Problems of social planning such as: what effect is exercised on the processes and structure of society by the scientific and technological revolution? are most important these days. This question raises numerous related issues: changes in the social, employment and education structure, the modifying effect on the superstructure and on institutions, changes in the structure of settlements and in ways of living, the new requirements of the management and planning of social processes in connection with scientific progress, the consequences for types of organization and communication, the redistribution of labour, etc. It is not mere chance that in the last ten years or so the interest of science was turned on itself, and science, having made itself the object of examination, undertook the science of science—a new complex domain of human knowledge.

Science, as a social phenomenon, and the scientific and technological revolutions as processes, undoubtedly have many aspects as objects of research. One of the most significant is the sociological one. The position of the sociology of science as a research field could be fixed at two places at least taxonomically: on the one hand, in the area of the science of science and, on the other, within the system of the field of research of sociology. By its position in the first, determined largely by the context, the sociology of science obtains an integrational affinity with those research trends and methods which examine science and scientific technological development from another non-sociological aspect. This aspect, 'from another side', may be philosophical, logical, psychological, informative, economic, historical, etc.

By its position in the second place, the sociology of science shows an even stronger integrational affinity with the problems of other areas of sociology and other sociological research directed to different problems. This shows that in this view, particular sciences do not have an *a priori* favoured and exclusively fixed place in the system of sciences. From a systematic point of view, the sociology of science is still within the general framework of sociology for it draws its methodological apparatus from the latter. As regards its relevance for social planning, the approach is a dual one, partly by the mediation of the science of science, partly as a kind of sociology within the framework of sociology as a whole.

The present study examines the role which is and can be played by the science of science in the planning of both the scientific and technological revolution and of social progress in a wider sense. A possible definition of the science of science is that it deals with the general structure of science, the ways and forms in which it functions and the dependence of its directions of development on other social phenomena and institutions. The goal of the science of science is the elaboration of those theoretical bases of the organizing, planning and directing of science which ensure the optimal pace and efficiency of the growth of science and also the application of its results in the interests of social progress.

As the definition shows, the science of science should be regarded as a complex theoretical and applied social science intended for prediction. Its most essential field of application can be seen in the forecasting of social, technological and scientific developments. But what are the possible applications of the science of science to forecasting? Because of both its underdevelopment and the fact that its outlines have only recently taken shape, it is difficult to separate two questions: i.e. the predictive functions of the science of science and those of science as such. If we refer either to the role played by science in social planning or to the scientific character of planning, we cannot be sure whether what is referred to is the systematic and deliberate use of the science of science or only the application of certain results of the various disciplines. Despite all these difficulties, an attempt will be made in this chapter to enumerate the possibilities for forecasting which the science of science offers.

Science as system and the science of science

Two notions have to be clearly distinguished, that of a science and that of a science the object of which is science. Science is a complex total and open system of social activities and achievements. The expression 'science as system' does not carry the same import as that of 'science as the system of knowledge'. 'Science as a system' is a wide, complex concept which covers the various elements of its structure, including scientific knowledge. These elements can naturally not be identified with each other. I, therefore, in this sense distinguish between scientific knowledge as an intellectual product and 'scientific production' that is the activity of scientific research. What I propose, however, is that the system of knowledge acquired through research activities be named not science, but more concretely and, in a more narrow sense, scientific knowledge. This naturally only includes knowledge systematized as theories; one must distinguish it from practical activities and empirical observations on the everyday level.

Scientific activity produces methodically obtained, confirmed, unceasingly multiplying and more perfect theoretical and objectified knowledge as the totality of a system of achievement. This activity at the same time gives shape to the system of the methods, processes and instruments of science. Scientific activity, therefore, has as its object the discovery, reflection, employment and prediction of the properties and the laws governing changes within natural, social and intellectual reality. It is an intellectually reflecting activity which formulates reality in concepts, theories, laws, theorems, algorithms, and other instrumentalities that help in carrying out its tasks.

An institutional aspect of the activity, which is in itself an objectified achievement, arises in the interests of optimum scientific research and its direction and organization. Institutionalization, accordingly, serves the planned direction of scientific activity. Scientific activity and its institutionalized form in addition find expression in connection with the organization of work and the division of labour. It is these that give shape to professional relations.

To sum up, the notion of 'science as a system' includes research, that is the activity which produces science, the necessary methodological and instrumental apparatus, and the sub-systems of institutionalized organizations and of professional relations within it which also produce the system of scientific knowledge. This analysis of the concept indicates its complex import. Its total nature derives from the fact that it is on a qualitatively higher level than the arithmetical sum of its partial systems. The system is open since it is continuously being extended and it constructs closer and closer ties with the systems of other social activities like production, politics and art. Science as a system is social since it is socially conditioned. It is a system of activities inasmuch as the peculiar and systematic epistemological apparatus of research lies at its very centre. It is a system of results and achievements inasmuch as scientific knowledge is accumulated as the product of this activity, but organizational and professional relations and methods are also realized in the form of results or achievements.

'Science as system' is the subject of the science of science. Nature is the object of individual scientific disciplines such as physics, chemistry or biology, or else it is a particular field of social or intellectual life. Science in the comprehensive sense, therefore, means neither the totality of such particular, specialist sciences, nor the science of science itself. The cumulative structure of science includes the sum of the particular sciences plus the science of science which examines 'science as a system'. The science of science thus is not a totality of scientific disciplines. It is more than that and more extensive, but also less and narrower. Its subject is the general structure of science, its mode of operation and forms, the way the rate and directions of development depend on other social phenomena and institutions. To that extent it is a broader notion since particular disciplines do not occupy themselves with such questions. It is narrower since it does not include those elements which can be found in particular sciences. Science, in the broadest possible sense, is the totality of particular sciences and the science of science. Taken in this sense, the science of science is one of the parts which make up science.

Precisely for that reason the science of science can become a direct force of production. This means that scientific creative work turns into the principal propelling factor of the forces of production. Just as what is achieved by particular disciplines can be applied in productive and social practice, knowledge relating to 'science as a system' can be applied in science policy, and in the direction of science and research activities. It is on this basis that the predictive functions of the science of science can be distinguished from those of the specialized scientific disciplines. In other words, while the science of science has an integral predictive function the various specialized scientific disciplines have differentiated predictive functions. In my view the ensemble of the functions of these two groups makes up the predictive functions of science.

The prognostical functions of the science of science

No claim is made, of course, that in each case the direct influence only of the science of science will be considered. Its indirect influence is certainly not negligible. We are faced with various phases of directness and indirectness, the comprehension of which can be of utmost importance from the standpoint of developing the science of science. This can be seen in the very fact that the essence of progress appears in the gradual transition from indirectness to directness. In this way the science of science cannot evade the validity of the thesis that science becomes a direct force of production. In my opinion the direct or indirect forecasting functions of the science of science can be discussed in the following way.

In our age most of the sciences have reached a stage of development in which, while preserving their empirical/collective, descriptive, comparative and systematizing/functions, they can envisage the discharging of new and, what is more, theoretical ones. Among these is the function of forecasting, together with that of explaining and generalizing. Modern sciences have proved increasingly useful for making forecasts. A paradoxical situation has, most regrettably, arisen in this case in which those sciences which directly serve social planning —the social sciences including the science of science—are now only in the empiric stage of development and are thus not able to discharge their predictive functions at the level of theory and laws. The most advanced sciences, on the other hand—mathematics, physics, biology and chemistry—are not called on when making forecasts. It is the empiric character of predictive sciences which even today makes their scientific character doubtful. This paradoxical situation—i.e. that the least developed sciences must take the longest view and moreover are required to make forecasts for the more developed ones—cannot yet be regarded as easily resolvable. It can be improved only by the consistent development of the social sciences, particularly the science of science. I shall not consider the epistemological roots of the problems involved in the numerous factors contributing to social questions or in the movement of the more developed systems according to the laws of probability, but in the course of research these must be taken into account.

The network model of science

In the traditional classificatory sectorial model the disciplines are often depicted as isolated, divergently growing, branches of a ramifying tree. Contrary to this, the network model of science, as proposed by J. D. Bernal, focuses on the integration of disciplines. The emphasis is laid not upon the detachment of the scientific attainments and methods, but on their concatenation. Network models have predictive value because they indicate the focal points of science, thus, exposing the system of manifold relations which can be developed by linking various experiences and methods. These models can be used in the forecasting of 'intersections' as well as in the prediction of new fields of knowledge and their 'catalysis' from the point of view of planning and organizing.

There are, however, different methods for forecasting the growth of particular sciences. The ontological method makes it possible for a science to enlarge its field of investigation. Epistemological aspects make it possible to forecast when a given field of research will reach a scientific, and hence more reliable, stage. The methodological system can be used for forecasting the manner in which the methods of various disciplines will merge into other research fields, and what the results will be. Functional analysis enables us to forecast the eventual time, extent and possibilities of the application and use of scientific achievements and research experience. Additionally it is the functional analysis of science which allow one to meet their organizational requirements. This in turn is equivalent to an assessment of their efficiency. The functional method makes possible not only mere classification of research fields according to disciplines, but also allows them to be integrated by the characteristic features of their structures into a complex body for doing research into given objectives. To give an example, I developed a method of training chemical engineers which consisted of problem spheres of varying composition that were functionally interconnected.

One can speak of the ontological functions of prediction in the sense of the expected directions research can be presumed to take and the possible ways the results may be applied, as regards the work to be done in understanding the existing natural, social and existential reality. The properties and states of affairs research is likely to discover in the future, given certain historical, environmental, spatio-temporal, etc. conditions, are investigated. Also, work can be undertaken, for instance on the forms (ways and mechanisms) taking shape in the scientific types of various concrete kinetics: mechanical cybernetics, reaction kinetics, nuclear kinetics. Similarly, the appearance of scientific types like ecology can be studied.

It is especially advantageous to compile network models of the current science system in addition to the 'science maps' of earlier periods of scientific development. Historical studies provide the opportunity for epistemological analysis of the dialectics of past and present, i.e. to the active and creative combination of history and prognostics. The 'science map' and science network model of a given period always contain the diverse elements and 'element complexes' on the gnosiological level such as disciplines, research trends, subdisciplines, etc. and their focal points.

Methodological prediction on the other hand refers to forecasting the way in which research methodologies elaborated in particular fields of research invade other fields of research and scientific disciplines; for example, the way in which mathematical or cybernetic methods conquer particular scientific disciplines.

The prediction of technical development and production

If we build up a theory of the forecasting of science within the framework of the science of science, this will raise future research planning to a higher level. This is an approach we should use even today when drawing up long-term plans. Hungary's fifteen-year scientific research plan is, for example, being drawn up on the basis of this approach. The planning of science on this basis further enables forecasting of technical development in different countries, but I shall not deal here with an analysis of this obvious interconnection.

It also seems superfluous to state that on the basis of scientific and technological forecasts the developing and planning of production can be predicted. At the same time, this indicates how erroneous the traditional thesis is that plans for scientific and technological research be deduced from the production plans of the economy. In Hungary, concurrently with the launching of the new system of economic management, essential changes were initiated in, for instance, the directing of technological research. The most important factor leading to these changes was the recognition that no technological development plan can take into consideration the size of the country alone, its level of technical development, the 'openness' of the economy and other constituents, because no such plan can be deduced direct from the national economic plan. It has been laid down in the Hungarian Socialist Workers' Party science policy directives that the fulfilment of national economic plans needs up-to-date technological expertise and sound research in each case. A proper survey enables us to envisage and determine what research has to be done in the country and what expertise has to be acquired in some other manner (cf. Science Policy Directives, 1969).

The main research trends that could and should take priority in Hungary between 1971 and 1985 were predicted from the four aspects just discussed. Using the ontological aspect, we made an analysis of objective features of the systems of natural and social reality. The gnosiological aspect was employed to predict the levels of scientific cognition feasible within a particular period. The methodological aspect predicted the extent to which the methods of the various fields of science penetrate one another. Finally, through the aspect of functional analysis, we gauged what had to be done about the organization of research. The predictable requirements of society and the economy with special regard to the organization of complex research themes were operationalized.

These considerations were accompanied by research on the selection of the main trends. Several different fields were identified: solid bodies, the mechanisms regulating vital processes, scientific administration, socialist enterprises and biologically active compounds. Additionally, the following research programmes were given priority: the aluminium industry, petrochemistry, measuring technology, complex light structure, building technology, human macro- and micro-environments, mechanical engineering technology, electronical spare parts, telecommunication, soil fertility, meat production and food diversification and processing.

The social factors conditioning science

The development of science takes place within society as a whole and is conditioned by many factors. Its rate and volume of productivity depend to a considerable degree on social conditions such as: the social structure, the given level of the labour structure, the culture and customs of the population, the standard of education of the labour force, the nature of the settlement-network, the standard of the infrastructure, demographic distribution, the level of development of the forces of production, the relations of production, etc. All of these have a great contribution to make to the general development of science within a country and to those problems in particular which are expressed on the basis of the needs and requirements of society and which call for scientific solutions by the research workers of the nation.

It is the forecasting of the subjects of what are called 'national' sciences which is specifically dependent on a concrete analysis of the actual conditions of the given society. The characteristic features of development in Hungary appear in, among other factors: the structure of 'quasi-development' of research. This means that from the point of view of the most important indices, the scientific research apparatus in Hungary can be compared with that of the most developed countries. However, the development of our research network remains illusory if we restrict our examination to the quantity and quality of the scientific achievements it produces. As a result of certain features of historical development no satisfying co-operation between research systems, education and production has been established and mutual cross-fertilization is also still lacking. The long-term development plan now being prepared is meant to transform the 'quasi developed' structure into a developed research and production structure. However, for this transformation it is necessary to undertake science of science research in combination with sociological, economic, organizational, historical, philosophical and psychological analysis, then it will be possible to promote the exploration of those factors which exert a positive or negative influence on scientific and technological development.

The effect of science upon social planning

Another group of forecasting possibilities can be found on the feedback side of the relations between science and society. While the social factors conditioning science have just been enumerated and their eventual future state was forecast, we now turn to the examination of the influence science exerts on these factors. Social planning can be named as the most essential function of the science of science. It expresses the synthesis of the various disciplines. Our times bear witness to the increasingly significant evolution of this function.

The types of social planning can be divided into three groups (cf. Kulcsár, 1969). In the first, we can forecast the social influence of the economic processes. The second involves the elimination or diminution of the disfunctional effects of the socio-economic processes through deliberate planning (e.g. parallel with the increase of the urbanization process, crime is also on the increase, which we are trying to reduce). The third type is social policy, also calling for highly co-ordinated planning. In the early days of the building of socialism, the main principle for dealing with social problems held that the character and direction of other basic social processes (e.g. the elimination of social inequalities) were automatically determined by the changing economic situation. Hence, the first socio-political measures had an *ad hoc* character. They were aimed at easing the sharpest tensions. Now the signs of development of a social policy constituting an integral part of long-term planning are increasingly present. Formerly, we did not take into consideration the social consequences of certain economic or socio-political decisions. Owing to the absence of social planning, these were spontaneous and often disfunctional; for instance, the policy of industrialization, which followed the Second World War, resulted in considerable demographic mobility. In the absence of a settlement policy and rational social policy, however, social tensions arose which were the negative concomitants of a fundamentally necessary social process. These must now be dealt with.

In the light of the considerations of the science of science it is obvious, of course, that economic planning holds a central place within social planning, since a social plan that contradicts economic processes results in the failure of the economic plan and may also become the source of serious social crises. It is one of the tasks of social planning to evaluate and influence the social consequences of planned and spontaneous economic processes. Social policy and regional planning are essential elements of social planning, but the planning of scientific and technological development should be added. Nevertheless, the relations between economic and social planning often conflict, e.g. a paradoxical situation may arise in which efforts directed to immediate economic efficiency may, in the long run, come into conflict with the expectations of social planning. To ease conflicts it seems necessary to elaborate on a scientific basis the idea of social efficiency while bearing in mind that not every measure that seems beneficial in the short term can be regarded as efficient in the long run.

From the outset the preparation and forecasting of settlements policy consequent on industrialization, the urbanization of villages, demographic policy, the educational system, regional planning, etc. require co-operation between the social sciences. The specific differences, characteristic social function, research organizing possibilities and attributes of the social sciences can, however, only be understood and deduced from a perception of the essence of science. This constitutes one of the responsibilities of the science of science. Science therefore also plays an indirectly predictive role by both stimulating the social function of the social sciences and delineating the nature of this function in social planning.

What I have in mind when I speak of the 'general essence' of science is principally the particular character of 'science as a system' which is examined by the science of science. When preparing social planning in a scholarly and scientific way, science policy must therefore, in the first place, bear in mind the general structure of science, the modes and forms of its operation, the directions of likely progress and development and connections with other social structures. I am inclined to argue that the proposition suggested by W. I. Thomas that if men accept a situation as real its consequences tend to become real, is not relevant to the analytical aspects of the predictive functions of the science of science. The notion of the self-fulfilling prophecy does not apply here, since the frequent and large deviations between forecasts and what actually happens suggest that reactions to the situation on the part of research workers, the organizers of science and the practitioners of science policy are often in vain. The objective features of the situation deflect implementation from intentions in a manner that is contrary to expectations. If the prediction is unsound, this will be unambiguously demonstrated. If the consequences are disfunctional, it is likely that a great many factors were not considered or could not be taken into account.

Regarding the interconnections between social planning and science, several points must be clearly distinguished. It is one thing to examine the social consequences of science (cf. Bóna and Farkas, 1970) and quite another to examine the means by which science may help to reveal the socio-economic processes which, on the one hand condition the development of science and technology, and on the other precondition the general manner of social development *per se*. It is still something else to examine what sort of scientific and non-scientific means we require for our intervention. We plan, through science itself, those effects of science aimed at social changes, but it is also necessary to investigate and apply factors which cannot be regarded as scientific. In this way we arrive at an elucidation of the roles of power and politics.

The relation between the science of science and politics

The science of science, as already indicated, does not only examine science but also makes an analysis of its links with other kinds of social activity. Thus, it tackles the place and interconnections between the political and scientific components of social planning. In referring to politics, we must be clearly aware of what we mean. Politics is a special power activity which cannot be regarded as a science, but the characteristic features of which can, since they are also socially conditioned, be understood through a scientific examination of these conditions, primarily by the sociology of politics (cf. Kulcsár and Farkas, 1969).

We must distinguish between the practice of science policy (strictly speaking, it is part not of science but of politics) and science policy which systematically orientates itself not by politics but, as a domain of political science, in terms of scientific theory. The identification of the practice and theory of science policy may often be a source of difficulties and misunderstandings. It is often the underdevelopment of science policy that is criticized instead of the backwardness of research into science policy We also eulogize science policy though it is worked out only as a theory. Furthermore, science policy can also be considered as both 'scientific politics' and the 'politics of science'. In the first case, we have to answer the question what effect science has on the formation of policies. In the second, we wish to know the manner in which politics influence the development of science. Either case requires, of course, scientific research. Accordingly, the science of science, examining the essence and connections of science with other social phenomena (e.g. with politics) is concerned with clearing up the following problems of prediction: how can science contribute to the making and adoption of optimal political decisions and attitudes. This does not mean, naturally, that co-operation with various special disciplines is not needed. On the contrary, it preconditions fruitful co-operation between them and epistemology.

From another point of view, we speak in Hungary of science policy as an integral part of state policy aimed at the acceleration of the process of building socialism with the help of science. On the basis of science policy considerations political bodies declare themselves on the following five questions: (1) the definition of the aims and trends of scientific research and of the proportional development of the various domains of research; (2) the determination of that part of national income which can be allocated to research objectives; (3) the proportional distribution of the research centres; (4) education of research scientists, promotion of both research scientists and research work, establishment of the necessary organizations and institutions, definition of the forms and means of the organizations which undertake and supervise research work; (5) the establishment of the directions and extent of the country's participation in the international division of labour (Lőrincz, 1970).

Hence, science policy comprises two factors: the determination of ideas related to the development of science as well as the means and methods ensuring the implementation of these. Science policy is the field in which all the predictive trends so far enumerated are combined, since the science hypotheses based on the immanent development of science, the technical and production development forecasts that can be deduced from them and the forecast of the role of scientific components in the gross social development are all combined and orientated towards a common aim within the framework of social policy. Through its complex aspects, the science of science promotes the elaboration of the components of science policy. It also provides forecasts in a wider sense for general politics because it co-ordinates not only the development of science, but also the related progress of all social factors. Forecasts of sector policies are also made on the basis of a scientific analysis and are imbued with an elaborate scientific attitude.

The prediction of the requirements of scientific-technical development

The central problem of the science of science is the forecasting of the social effects of the scientific and technological revolutions. At the same time, it is precisely this achievement by which it can contribute directly to science policy and to social planning.

The scientific and technological revolutions and the investigation of their claims may be regarded as the fundamental road of the forces of production leading to the completion of the material and technical bases of socialism and communism. As to working tools (this means full automation as to the relations between man and machine) it expresses the supremacy of man, nature as a productive force of science and associated with it, the comparative reduction of direct work by the production potential, the high level of professional and general skills and their overall diffusion, and a process leading to the highly developed organization in its social and even international aspects. Its social preconditions can be summed up as: the formation of the socio-economic and cultural environment and public opinion in which acquisition and application of new knowledge and spontaneously active work become a fundamental requirement of both the voluntary bodies and the working collectives while the system of directing society keeps on growing, as does its organization and general policy. It is this that lays down the institutional system of the methodological development of sciences and of the acquisition of knowledge as well as the material and technical base for the practical application of results. It also presents a value system and a consumption pattern allowing for the proper material and moral appreciation of the effectiveness of socially useful work.

On the evidence of a study undertaken by economists, sociologists, psychologists, philosophers and other professional people (National Committee for Technological Development, 1970), one can state that today in Hungary the political and economic information available and the resulting public opinion and behaviour may, in coming decades, undermine planned rapid technical progress if general knowledge is not significantly increased by the deliberate and organized use of all the educational resources including adult education. It is, therefore, not only the application of scientific achievements in production which should be accelerated, but their eventual use and penetration into the consciousness and activity of the workers. We do not know a great deal, however, about the laws of the diffusion of scientific knowledge among the public. It is my view that this diffusion also constitutes one of the essential fields of research of the science of science. In the interest of future planning of changes in the social structure, one must start from the present.

This is true not only of the material, technical and economic nature of things, but also of the structure of consciousness, culture and ways of living, and the standards of apperception as well as attitudes and behaviours. This underscores the close relationship between science policy and social policy and the general importance of specialized sociological studies to both.

References

Bóna, Ervin and Farkas, János (1970), 'Some contradictions in the present state and structure of science', *Magyar Tudomany*, No. 6 (in Hungarian), Budapest.

Kulcsár, Kálmán (1969), 'Social efficiency and social planning', lecture at the Balatonfüred International Sociological Conference, 9–12 September, 1969.

Kulcsár, Kálmán and Farkas, János (1969), 'On the possibilities of political sciences', *Pártélet*, No. 7, Budapest.

Lőrincz, Lajos (1970), 'Some theoretical questions of science policies' in Bóna, Farkas, Lőrincz, Klar and Paczolay, *Some Theoretical Questions of Science, Essays*, Budapest: Akademiai Kiadó.

National Committee for Technological Development (1970), *The Social Preconditions and Possible Effects of the Success of the Scientific-technical Revolution in Hungary in the next 15–20 years* (in Hungarian), Budapest.

Science Policy Directives of the Hungarian Socialist Workers' Party (1969) *Társadalmi Szemle*, Nos 7–8 (in Hungarian).

14

Bureaucratic trends in the organization and institutionalization of scientific activity

Zdislaw Kowalewski

Being aware of the emotional overtones and the tendency for pejorative interpretation connected with the term bureaucracy, the author wishes to discuss two aspects of its meaning:

(1) postulates or instructions in which the notion is defined as a trend towards rational organization of man's collective activity based on defined principles,

(2) empirically observed deviations of the postulated bureaucratic conceptions, the real functioning of a scientific research organization. The confrontation of the postulated and the real models with the organization of scientific activities is aimed at answering the question: whether and to what extent the postulates of bureaucratic organizations can be conducive to the improvement of scientific activity. Our interests are restricted to the contemporary organizational features of one country—present day Poland.

The postulated model of bureaucracy

Those who, like Saint Simon or Max Weber, saw the science of society as a means of shaping reality were interested not only in the analysis of existing models of social organizations, but also in postulating models, ideal types, which could play a positive and active part in the formation of reality. Modes of the organization of social life were born spontaneously. They preceded the trend of rational formation of those models. Due to that, Auguste Comte saw the basic cognitive problems of the social order formed according to 'natural laws'. In the existing types of social organizations, he observed a variety formed under the influence of forces not fully controlled by the intellect of organizers.

The concept of bureaucracy presented by Max Weber as a model for the rational organization of human co-operation, an efficient form for organizing institutional activity viewed in contemporary associations, can be stressed as follows:

(1) Variation and expansion of specialization as defined by means of objective criteria of individual qualifications. The prestige of particular posts historically connected with conferment of superior posts to persons of social merit, became synonymous with the position. Sometimes succession or an appointment to a position became synonymous with the acquisition of specializations and proper qualifications.

(2) Differentiation of the range of competence and responsibility is subordinated to the hierarchic arrangement of posts: the higher position in the internal division of labour the wider the range of responsibility; the higher position the higher the qualifications and the range of competence.

(3) Documentation of all procedural elements is done by explicit and lasting symbols intelligible to the members and being easily controlled. This leads to overgrowth of paper work, office work, documentation work, and posts of controlling and auxiliary functions.

(4) Aims and duties of an association formulated into statutes, regulations, etc. defining ways of acting in order to achieve its goals. Independently of those general formulations there exists the necessity of a systematic actualisation of norms by issuing detailed rules and regulations to mark the correct behaviours leading to the realisation of aims. Institutional rules and prohibitions define in substance the subject of the activity, but they also include instructions concerning the attitude adopted towards other individuals that take part in the collective organizational enterprise.

(5) The concept of man entangled in the processes of organizational co-operation assumes the emancipation of the role of an official from other individual social roles. Interhuman relations are based on a rational link of common concern for the subject of activities. Personal emotions, individual desires, inclinations, habits, etc. cannot influence the behaviour connected with execution of the function of an employee or member of the organization. The relations between individual

members of the organizational systems are normalized by formally recorded norms and protected with sanctions.

The postulated model of bureaucracy recapitulated in these five points is sometimes identified with the concept of formal organizations based on 'scientific principles' in contrast to informal and spontaneous models of institutional behaviours.

Its particular elements deal not only with the historically formed organization of power, administrative and military structures, but mostly with modern industrial organizational systems. Each type of institutionalized human activity adopted the bureaucratic model—with a few differences in the ecclesiastical, military and educational systems of organizations.

Bureaucratic systems of organizational behaviours

The pattern of behaviours in modern associations is a subject of specialized studies and can be treated in this chapter as a frame of reference for descriptive features of organizational norms. The dysfunctions of a bureaucratic model for organizational behaviours can be stressed as follows:

(1) Specialization and individuation of particular functions in an organizational system leads to segmentation and quantitative expansion of participants in these activities, so that the problem of coordination of their joint action emerges more and more as regulative functions. On the other hand, definition of the range of qualifications and their submission to the determined current targets makes the personnel less efficient in new situations which have not been predicted by stabilized models. Simultaneously, the range of specialization and qualification of various organizational functions is extremely difficult to define accurately. Respective specialities frequently cannot find a 'common language'—the joint system of values that would enable a concurrent interpretation of common goals and their institutional views. In those cases, the functioning of the system becomes the most important and elucidative value of actions. But identification of associational aims with the socio-technical very often leads to a general alienation and external forces directed to or against the association may have a functional unity. The symbolic features of the highest qualifications often play a more remarkable role as the real profits of specializations.

(2) The organizational hierarchy based on rational prerequisites of

the range of qualifications, competence and responsibility does not eliminate the influence of other norms and values that transfer the official criteria of hierarchy and competence. The differentiation of the range of competence connected with a great number of those participating in the collective activity leads at the same time to a narrow range of responsibility. Execution of the targets sometimes becomes connected with violation of one's own range of competence. It may be contradictory to the existing division of work and responsibility; the person who makes the decision is not always able to secure its realization. In those cases, 'collective decisions' which exempt the individual from responsibility are taken. Lack of individual responsibility from the decision-making process shatters the postulate of efficient action.

(3) Documentation of acts connected with activity must employ elaborate schemes which can be justified by those who analyse or supervise the activity. Nevertheless, some activities require documentations which are extremely complicated or even impossible in certain circumstances, e.g. when they deal with the process of thinking, learning, experimentation, construction, etc. Documentation of many activities can be so tedious and laborious that it consumes more time than their execution.

Certain specialized activities are subordinated to the technique of utilizing symbolic documentation which can be understood by specially qualified personnel. Use of documentation standards legible to laymen demands auxiliary activities which seem redundant or even unreliable to specialists. That happens when central supervisory posts are not able to take into account the existence of separate techniques of action in all domains of social life. Supervision over performance then becomes an isolated and useless activity. When treated as an impulse for efficient execution of tasks, supervision often brings about deviations within the sphere of documentation: it leads to the so-called 'double book-keeping', one for supervision and another for efficient acting.

(4) Norms of the behaviour defined in statutes, regulations and instructions may be interpreted in many ways. The tendency of their increasing specification leads to such a great quantitative increase of regulations and instructions and to get acquainted with them becomes practically impossible for the performers. The objective situation and functioning of a given institutional system, its cultural aspects, people and instruments entangled in the processes, always deviate from the models employed by the legislator. This is why norms and

laws do not constitute sufficient factors of evolution and interpretation of reality. The interpretation is usually performed by the superior who passes opinions on acts of the subordinates and makes him responsible for them. At the same time, subordinates pass opinions on superior's acts and learn a lesson of an autodidactic and social discrepancy.

There is often a tenuous bond between definite concepts of anticipated human nature and the real behaviours and systems of values of the actors.

(5) A similar problem appears while examining an assumption that participation in an organization is connected with the possibility of a full and entire identification with the role of an official; an assumption is made that there is a possibility of excluding oneself from other roles which are to be performed within this and other social institutions. Sociological analyses of the functioning of modern organizations show that every employee has to perform many social roles within the same organizational structure. Therefore, in a Polish state's ownership system an employee is obliged not only to perform a role of the functionary, but mainly of a co-manager, an expert, a member of politicalized associational groups, a colleague, as well as the role of a politically conscious and active member of the local administrative systems.

Sociologists in their empirical studies on organizational systems have found that the role of a functionary is not interpreted identically by all employees engaged in the process of participation. The chances for participation and interpretation of social roles are different: they depend on the position acquired in work; expectation in these aspects vary in industrial, technological, educational and research organizations.

The postulated model of bureaucracy is based on an optimistic assumption that positive co-operation is the dominant mode of human relations. Disputes and conflicts constitute only a marginal problem, a disfunctional aspect to be eliminated.

Some types of organized human activity are based mainly on an assumed 'negative co-operation': those are mainly sport, military and political ones. Though their internal organization is based on the principle of positive co-operation the necessity of functioning posits the emergence of adversaries or enemies with whom relations are based on the principle of competitions, contests, struggles. Dissemination of models of the 'negative co-operation' which are characteristic of those organized forms of human activity is reflected in such phrases as front, vanguard, combat, mobilization of reserves, etc. The 'negative co-operation' occurs mainly in situations where distinct and contradictory aims exist. Assuming that the existence of positive co-operation is connected with the acceptation of common institutional goals and the negative with different interpretations or different modes for achieving them, the studies of real patterns of behaviours become very important. Are the aims of a state's enterprise and a state's association identical with those of their participants and workers ? Do particular elements of organized activity have common or distinct aims ? What is the range of common and distinct interests ? Empirical studies answer those questions. The results of those studies permit one to state that the 'negative co-operation' seems to be as common as the positive. Very often it is treated as a method of work that leads to common societal goals.

The postulated model of bureaucracy and scientific activity

The individuation of scientific activity from other forms of institutionalized human behaviours becomes very difficult when, on the one hand, the growing social prestige of science promotes the use of that word to describe all activities leading to cognition and, on the other hand, when more and more groups interested, to a certain degree, in that activity are created. To achieve some degree of unambiguity it is necessary to note distinct features of such synonymous terms as knowledge, wisdom, techniques, technology, methodology and science. Let us remember the distinction made by Aristotle between *δόξα*, *ὑπόληψις* and *ἐπιστήμη*, between individual opinion, social judgments and verifiable scientific knowledge. Knowledge constitutes the condition for any rational activity. It is linked with each social role performed by man. There are different types of knowledge necessary for each kind of institutionalized human behaviour and only the historically emerged methodological patterns of theoretical and experimental-empirical approaches towards cognitive goals received the name of scientific ones.

The limits between non scientific and scientific cognitive activity are not easy to define precisely. Therefore, in practice, the institutional interpretation appears most explicit. Nevertheless, many simplifications occur because the criteria of identification of some fields of scientific activities are different in various societies, similar to the variability of criteria in the same society in different historical periods.

Teleological knowledge for practical aims is usually called technology. The notion of technology is closely connected with the notion of instrumentalization and that is why it has different meanings in industrial laboratories and social praxis and theories. Technology treated as scientific approach towards practical targets, utilization of methodological achievements for cognitive activities in industry, economy or politics, became in the twentieth century a large-scale institutionalized scientific activity. It was traditionally assumed that scientific activity is engaged in search of the answer to 'why', in causal explanations of phenomena and facts, whereas an answer for 'how' in the teleological points of view constituted the characteristic feature of engineering and construction, technological endeavours, and practical aims. The evaluation of scientific work is not free of this historical tradition of thinking. Historically formed scientific types of institutionalized activities have definite connections with non scientific institutions. Up to this day, we can recognize different types of institutionalized patterns of scientific behaviours linked with religion, education, arts and the economy.

The first one has in currently institutionalized patterns of thinking and scientific behaviour tendencies of theoretical instructions, ideological interpretations and philosophical generalizations in different types of scientific communities all around the world. For many politicians, educators and clergies, this type of institutionalized scientific pattern of behaviours is treated as fundamental and highly valued.

The second type of institutionalized scientific activity is connected with the didactic and educational aims of society. Very well known throughout civilizations, it is still the most popular form of institutionalized scientific behaviour. The model of institionalized scientific activities created by nineteenth century German universities is still alive and prominent. Many believe that research activity performed without any links with didactic activity is useless; it does not help intellectual and research development.

The third type of institutionalized scientific activities formed by artisans, artists and experimenters mainly of the Renaissance epoch, gave new patterns for institutionalized scientific behaviours and scientific communities. When connected with the first pattern of institutionalization they created formal scientific associations known under the name of Societies and Academies.

The fourth type of scientific institutionalization was developed by the capitalist system joining the entrepreneurs' needs for economic success with the social demands for innovations in production and products. Nineteenth century industrial laboratories in Europe, then in the United States, produced new patterns of scientific-technological behaviours which dominate the current quantitative data about scientific institutions and scientific workers.

Each type of institutionalized behaviour is connected with different systems of values, different types of instrumentalization and different types of social institutions. The question about the unity of science has not only its methodological and epistemological roots but the social too—connected with the institutionalized patterns of thinking and acting. The idea of the unity of science connected with its ethological and epistemological cognitive structure leads to the stereotyped models of research organization used by politicians, educators, managers or economists. Standardization of organizational structures cannot give the best opportunities either for development of a scientist's personality or for the development of science and the efficient results expected by a sponsor.

Therefore, the question, whether a chance exists for working out an organizational model adequate for creative scientific activities, remains pertinent. An examination of the mentioned values of the postulated bureaucratic model is still required.

The rational organization of activity based on the postulated bureaucratic model has to formulate the internal division of work connected with specialized performances submitted to the teleological structure of the institution. For research activity, the teleological structure is fixed, first of all, by cognitive and methodological features regardless of the functions fixed by its organizers, whether they want it to be an educative, economic, technical or political one. Historically formed types of specialization in scientific activity are differentiated not only as regards the method applied and subject of research and the system of knowledge to which the activity is submitted, but also by the existing instrumentalization, cultural heritage and attitudes of the research worker.

Independently of what scientific discipline they represent and in what type of association the different types of scholar find adequate conditions for their work, they will interpret differently the teleological structures of their workshop. The definition of various types of specialists becomes particularly difficult as far as the postulates of the bureaucratic model are concerned, when beside the epistemological and cultural differentiation already mentioned, we consider the differentiation connected with social sponsorship and type of associations organizing the work. The internal division of work may require an advanced division of performances. This division may be based on the temporal sequence of the research process or it may be restricted to two fundamental functions: managerial and auxiliary as in the majority of traditionally organized types of handicraft, workshops, artistic, medical and scientific work. The organization of didactic-research work places is, in principle, based on the differentiation of functions of an 'independent research worker' and an 'auxiliary research worker'. Additionally, there are 'engineering-technical' workers as well as technical and administrative workers according to the model of organizational structure. In the present organization of work in Poland there are all these categories of 'research workers', but that does not mean that the actual names or titles correspond with the same functions in particular types of institutions. The actual research, scientific role adopted depends on the task, the qualifications, instrumentalization and cultural conditions available for an anticipated function. If the system of salaries gives some advantages for certain categories of worker, the structure of qualifications becomes deformed and ambiguous. Increasing differentiation of organizational function in research activities according to the bureaucratic model changes the general cultural structure of scientific work and cannot be neutral for research efficiency.

But bureaucracy in scientific activity does not only consist in the differentiation of particular auxiliary specializations directly connected with the process of research, but also with the differentiation of administrative functions connected with the documentation requirements of the activity, not according to methodological instructions of a scientific discipline, but in accordance with the standards of administrative non-scientific institutional patterns; the economic, financial, industrial, political ones. Administrative employees engaged in scientific activity do not always represent the role of professional clerks with typical bureaucratic features; in small research units their roles are similar to auxiliary and technical functions supporting scientific activity. But, in central organizations where their numbers exceed that of research workers, e.g. in ministries supervising universities, in large department institutes, in the Polish Academy of Science, very often they become not the technical aid for scientific activities, but a group for management and control. Their main concerns are connected with problems of financing, personnel policy, administrative planning, administrative reporting, and co-ordination of activities; not with actual research, as they do not understand it.

The predominance of regulating and auxiliary concerns over the fundamental does not create a type of institutionalization suitable for scientific activity.

The range of competence and responsibility in scientific activity is marked, first of all, by specialization which is different from administrative specialization. The range of methodological responsibility differs very much from the range of organizational responsibility in the complex, interdisciplinary and multi-directional structures of scientific institutions. The manager and director of a research laboratory who was formerly in charge of, for instance, the mechanics section, may be recognized as responsible only for those sections that carry out activities concerned with the methodology known to him. It is more difficult for him to formulate problems and consistently analyse the research work of, let's say, physical chemistry or biochemistry units.

The hierarchy which develops as a result of scientific activity is connected with experience based on intellectual contact—not on organizational supremacy. Very often interpreted as 'scientific supremacy', it constitutes an obstacle in the organization of cognitive performances. According to the scientific community, every independent scientist decides the range of his interests and defines his research strategy. Definitions of a field of research need not always be identical. Sometimes there are a greater variety of research initiatives in small units; but if the sponsor has little tolerance then only big research laboratories provide an opportunity for non-standardized creative work. If an institute was created as a result of individual efforts, individual organizational links constitute a compact system. The teleological structure of such an institute may be called monolithic. The contrary situation occurs if a scientific institute, constituted of various groups of interests, disciplines and scientific communities, is created as a result of an administrative decision or personal endeavours to be a scientific director. In that situation, managerial posts spring up and an administrative hierarchy of values and behaviours replaces the institutionalized scientific cultural structure. In order to support their authority, managers increase the number of administrative employees and seem not to notice the differences between formal, or organizational, superiority and intellectual authority. Activities in such an institute resolve themselves, mainly, into the organization of auxiliary activities such as the provision of equipment, conferences and documentation popularization. Though they seem to be of monostructural character, other big scientific institutes may constitute federations of autonomous establishments in spite of the formal legal status and the existence of only one top managerial post. The role of management and administration appears similar to the organizational function performed by federative departments.

The mode of scientific documentation is usually determined by the characteristic method of a research field. It differs according to the methodological approaches and research techniques. The basic elements of an investigator's work-place comprise instruments for the documentation of ideas, plans, schemes, suggestions, fragmentary results, sketches, information obtained, etc. One could say that the scientific worker's individual work-place resolves itself mainly into the documentation of various kinds of information. No homogeneous standards have been elaborated as far as that issue is concerned, but studies enable the specification of heterogeneous forms of techniques of work. Postulates concerning the documentation of activities which are included in the model of bureaucratic organization are related to an effort to supervise those activities. Supervision aims at the recognition of stages of realization of targets. In the case of an economic activity where the result may be defined by means of financial indices, documentation may be helpful in explaining the causes or the effects of particular results of that activity, where as in the case of research activity this is not the case then documentation of distinct performances may not always play that role.

The norms which regulate the behaviours of members of a working group in the postulated model of bureaucratic organization are based on a system of stabilized values—a system of statements of dogmatic character. The norms of research processes come from epistemological structures and methodological patterns. The ethological norms of scientific acting, mainly scepticism and communism, are strange for non scientific institutions.

One can say that the norms of universalism, communism, disinterestedness and organized scepticism are absent in the cultural structures of an administrative or industrial institutionalization. Current trends towards the identification of scientific workers with functionaries of research or educative institutes instead with the 'institution of science' are the source of many difficulties for creative activity. The bureaucratic mode of organization creates different social roles for the same research worker, as a functionary specialist, co-worker, supervisor, staff-member, ideologist; but biographies of scholars show that when the investigator's role is treated as the most important it very often renders impossible a serious execution of other social roles.

Conclusion

The existing organizational models universalized by economists and administration workers are based on stereotypes inadequate for scientific activity. The different types of historically formed institutionalizations and different epistemological structures necessitate their own norms of organized behaving. But, there are some common features of the cultural structure underlying scientific behaviours which can be treated as fundamental for rational organization. These include: the teleologic, epistemological, ethological and technological elements. The first is connected with the social criteria and goals which allocate priorities for different research activities; the second with cognitive values which determine the nature of the scientific work; the third with specific social values and norms such as communism, disinterestedness, scepticism, forming the organized human activity and the technological, refers to the kind of instrumentalization and techniques required for the specified research work.

There is a need for the design and elaboration of different organizational models for scientific activities institutionalized historically in the form of:

(1) international scientific communities of different specializations,

(2) national culturally and teleologically determined groups of activities connected with didactic educative institutes, economic, industrial, political, etc.,

(3) international institutes for general research co-operation,

(4) national and international institutes for the dissemination of knowledge and utilization of scientific and technological achievements.

The improvement of scientific workers' performance and of their methodological culture should be based on the epistemological and cultural structure of science treated as an intercivilizational historical institution.

Index